The 2011 Fukushima Nuclear Power Plant Accident

Related titles

Fukushima Accident: Radioactivity Impact on the Environment
(ISBN 978-0-12408-132-1)

Radioactive waste management and contaminated site clean-up: Processes, technologies and international experience
(ISBN 978-0-85709-435-3)

Handbook of small modular nuclear reactors
(ISBN 978-0-85709-851-1)

Managing nuclear projects: A comprehensive management resource
(ISBN 978-0-85709-591-6)

Woodhead Publishing Series in Energy: Number 73

The 2011 Fukushima Nuclear Power Plant Accident

How and Why it Happened

Yotaro Hatamura, Seiji Abe, Masao Fuchigami and Naoto Kasahara

Translated by *Kenji Iino*

ELSEVIER

AMSTERDAM • BOSTON • CAMBRIDGE • HEIDELBERG
LONDON • NEW YORK • OXFORD • PARIS • SAN DIEGO
SAN FRANCISCO • SINGAPORE • SYDNEY • TOKYO
Woodhead Publishing is an imprint of Elsevier

WP WOODHEAD PUBLISHING

Woodhead Publishing is an imprint of Elsevier
80 High Street, Sawston, Cambridge, CB22 3HJ, UK
225 Wyman Street, Waltham, MA 02451, USA
Langford Lane, Kidlington, OX5 1GB, UK

British Library Cataloguing-in-Publication Data
A catalogue record for this book is available from the British Library

Library of Congress Control Number: 2014952343

ISBN 978-0-08-100118-9 (print)
ISBN 978-0-08-100132-5 (online)

For information on all Woodhead Publishing publications visit our website at http://store.elsevier.com

Typeset by SPi Global
www.spi-global.com
Printed and bound in the United Kingdom

Contents

List of authors

S. Abe Kansai University, Osaka, Japan

M. Fuchigami Komatsu Ltd, Tokyo, Japan

Y. Hatamura Hatamura Institute for the Advancement of Technology, Tokyo, Japan

N. Kasahara The University of Tokyo, Tokyo, Japan

Woodhead Publishing Series in Energy

1 **Generating power at high efficiency: Combined cycle technology for sustainable energy production**
Eric Jeffs
2 **Advanced separation techniques for nuclear fuel reprocessing and radioactive waste treatment**
Edited by Kenneth L. Nash and Gregg J. Lumetta
3 **Bioalcohol production: Biochemical conversion of lignocellulosic biomass**
Edited by Keith W. Waldron
4 **Understanding and mitigating ageing in nuclear power plants: Materials and operational aspects of plant life management (PLiM)**
Edited by Philip G. Tipping
5 **Advanced power plant materials, design and technology**
Edited by Dermot Roddy
6 **Stand-alone and hybrid wind energy systems: Technology, energy storage and applications**
Edited by John K. Kaldellis
7 **Biodiesel science and technology: From soil to oil**
Jan C. J. Bart, Natale Palmeri and Stefano Cavallaro
8 **Developments and innovation in carbon dioxide (CO_2) capture and storage technology Volume 1: Carbon dioxide (CO_2) capture, transport and industrial applications**
Edited by M. Mercedes Maroto-Valer
9 **Geological repository systems for safe disposal of spent nuclear fuels and radioactive waste**
Edited by Joonhong Ahn and Michael J. Apted
10 **Wind energy systems: Optimising design and construction for safe and reliable operation**
Edited by John D. Sørensen and Jens N. Sørensen
11 **Solid oxide fuel cell technology: Principles, performance and operations**
Kevin Huang and John Bannister Goodenough
12 **Handbook of advanced radioactive waste conditioning technologies**
Edited by Michael I. Ojovan
13 **Membranes for clean and renewable power applications**
Edited by Annarosa Gugliuzza and Angelo Basile
14 **Materials for energy efficiency and thermal comfort in buildings**
Edited by Matthew R. Hall
15 **Handbook of biofuels production: Processes and technologies**
Edited by Rafael Luque, Juan Campelo and James Clark

16 **Developments and innovation in carbon dioxide (CO_2) capture and storage technology Volume 2: Carbon dioxide (CO_2) storage and utilisation**
Edited by M. Mercedes Maroto-Valer
17 **Oxy-fuel combustion for power generation and carbon dioxide (CO_2) capture**
Edited by Ligang Zheng
18 **Small and micro combined heat and power (CHP) systems: Advanced design, performance, materials and applications**
Edited by Robert Beith
19 **Advances in clean hydrocarbon fuel processing: Science and technology**
Edited by M. Rashid Khan
20 **Modern gas turbine systems: High efficiency, low emission, fuel flexible power generation**
Edited by Peter Jansohn
21 **Concentrating solar power technology: Principles, developments and applications**
Edited by Keith Lovegrove and Wes Stein
22 **Nuclear corrosion science and engineering**
Edited by Damien Féron
23 **Power plant life management and performance improvement**
Edited by John E. Oakey
24 **Electrical drives for direct drive renewable energy systems**
Edited by Markus Mueller and Henk Polinder
25 **Advanced membrane science and technology for sustainable energy and environmental applications**
Edited by Angelo Basile and Suzana Pereira Nunes
26 **Irradiation embrittlement of reactor pressure vessels (RPVs) in nuclear power plants**
Edited by Naoki Soneda
27 **High temperature superconductors (HTS) for energy applications**
Edited by Ziad Melhem
28 **Infrastructure and methodologies for the justification of nuclear power programmes**
Edited by Agustín Alonso
29 **Waste to energy conversion technology**
Edited by Naomi B. Klinghoffer and Marco J. Castaldi
30 **Polymer electrolyte membrane and direct methanol fuel cell technology Volume 1: Fundamentals and performance of low temperature fuel cells**
Edited by Christoph Hartnig and Christina Roth
31 **Polymer electrolyte membrane and direct methanol fuel cell technology Volume 2: *In situ* characterization techniques for low temperature fuel cells**
Edited by Christoph Hartnig and Christina Roth
32 **Combined cycle systems for near-zero emission power generation**
Edited by Ashok D. Rao
33 **Modern earth buildings: Materials, engineering, construction and applications**
Edited by Matthew R. Hall, Rick Lindsay and Meror Krayenhoff
34 **Metropolitan sustainability: Understanding and improving the urban environment**
Edited by Frank Zeman
35 **Functional materials for sustainable energy applications**
Edited by John A. Kilner, Stephen J. Skinner, Stuart J. C. Irvine and Peter P. Edwards
36 **Nuclear decommissioning: Planning, execution and international experience**
Edited by Michele Laraia

37 **Nuclear fuel cycle science and engineering**
Edited by Ian Crossland
38 **Electricity transmission, distribution and storage systems**
Edited by Ziad Melhem
39 **Advances in biodiesel production: Processes and technologies**
Edited by Rafael Luque and Juan A. Melero
40 **Biomass combustion science, technology and engineering**
Edited by Lasse Rosendahl
41 **Ultra-supercritical coal power plants: Materials, technologies and optimisation**
Edited by Dongke Zhang
42 **Radionuclide behaviour in the natural environment: Science, implications and lessons for the nuclear industry**
Edited by Christophe Poinssot and Horst Geckeis
43 **Calcium and chemical looping technology for power generation and carbon dioxide (CO_2) capture: Solid oxygen- and CO_2-carriers**
Paul Fennell and E. J. Anthony
44 **Materials' ageing and degradation in light water reactors: Mechanisms, and management**
Edited by K. L. Murty
45 **Structural alloys for power plants: Operational challenges and high-temperature materials**
Edited by Amir Shirzadi and Susan Jackson
46 **Biolubricants: Science and technology**
Jan C. J. Bart, Emanuele Gucciardi and Stefano Cavallaro
47 **Advances in wind turbine blade design and materials**
Edited by Povl Brøndsted and Rogier P. L. Nijssen
48 **Radioactive waste management and contaminated site clean-up: Processes, technologies and international experience**
Edited by William E. Lee, Michael I. Ojovan, Carol M. Jantzen
49 **Probabilistic safety assessment for optimum nuclear power plant life management (PLiM): Theory and application of reliability analysis methods for major power plant components**
Gennadij V. Arkadov, Alexander F. Getman and Andrei N. Rodionov
50 **The coal handbook: Towards cleaner production Volume 1: Coal production**
Edited by Dave Osborne
51 **The coal handbook: Towards cleaner production Volume 2: Coal utilisation**
Edited by Dave Osborne
52 **The biogas handbook: Science, production and applications**
Edited by Arthur Wellinger, Jerry Murphy and David Baxter
53 **Advances in biorefineries: Biomass and waste supply chain exploitation**
Edited by Keith Waldron
54 **Geological storage of carbon dioxide (CO_2): Geoscience, technologies, environmental aspects and legal frameworks**
Edited by Jon Gluyas and Simon Mathias
55 **Handbook of membrane reactors Volume 1: Fundamental materials science, design and optimisation**
Edited by Angelo Basile
56 **Handbook of membrane reactors Volume 2: Reactor types and industrial applications**
Edited by Angelo Basile

57 **Alternative fuels and advanced vehicle technologies for improved environmental performance: Towards zero carbon transportation**
Edited by Richard Folkson
58 **Handbook of microalgal bioprocess engineering**
Christopher Lan and Bei Wang
59 **Fluidized bed technologies for near-zero emission combustion and gasification**
Edited by Fabrizio Scala
60 **Managing nuclear projects: A comprehensive management resource**
Edited by Jas Devgun
61 **Handbook of Process Integration (PI): Minimisation of energy and water use, waste and emissions**
Edited by Jiří J. Klemeš
62 **Coal power plant materials and life assessment**
Edited by Ahmed Shibli
63 **Advances in hydrogen production, storage and distribution**
Edited by Ahmed Basile and Adolfo Iulianelli
64 **Handbook of small modular nuclear reactors**
Edited by Mario D. Carelli and Dan T. Ingersoll
65 **Superconductors in the power grid: Materials and applications**
Edited by Christopher Rey
66 **Advances in thermal energy storage systems: Methods and applications**
Edited by Luisa F. Cabeza
67 **Advances in batteries for medium and large-scale energy storage**
Edited by Chris Menictas, Maria Skyllas-Kazacos and Lim Tuti Mariana
68 **Palladium membrane technology for hydrogen production, carbon capture and other applications**
Edited by Aggelos Doukelis, Kyriakos Panopoulos, Antonios Koumanakos and Emmanouil Kakaras
69 **Gasification for synthetic fuel production: Fundamentals, processes and applications**
Edited by Rafael Luque and James G. Speight
70 **Renewable heating and cooling: Technologies and applications**
Edited by Gerhard Stryi-Hipp
71 **Environmental remediation and restoration of contaminated nuclear and NORM sites**
Edited by Leo van Velzen
72 **Eco-friendly innovation in electricity networks**
Edited by Jean-Luc Bessede
73 **The 2011 Fukushima nuclear power plant accident: How and why it happened**
Yotaro Hatamura, Seiji Abe, Masao Fuchigami and Naoto Kasahara. Translated by Kenji Iino

Introduction

On March 11, 2011, The Great East Japan Earthquake occurred, and the tsunami waves it induced triggered an unprecedented accident at the Tokyo Electric Power Company (TEPCO)-owned Fukushima-1 (Daiichi) Nuclear Power Plant (NPP). Three and a half years have already passed since then. Great efforts by those involved with recovery have placed the reactors in cold shutdown; however, the problems of contaminated water, over 130,000 evacuees (160,000 immediately after the accident) still kept out of their homes, and decontamination still continue.

On May 24, 2011, the Cabinet Council decided to establish the Investigation Committee on the Accident at the Fukushima Nuclear Power Stations. Its assignment was, for Fukushima-1 and Fukushima-2, to "conduct multi-faceted research and verification to identify the cause from a neutral standpoint of the people of Japan and propose policies to stop the damages from escalating and prevent similar accidents from repeating [1]." Naoto Kan, the Prime Minister at the time, appointed one of the authors of this book, Yotaro Hatamura, to chair the committee. Seiji Abe and Masao Fuchigami were appointed by Hatamura to serve as technical consultants to the committee.

After its first meeting on June 7, 2011, the committee visited the sites of Fukushima-1 and -2 NPPs for inspection and interviewed a total of 772 people: the Prime Minister and others from the central government, heads of affected local governments, residents and evacuees of the area, and others. The committee published its Interim Report on December 26, 2011 [1], and the Final Report on July 23, 2012 [2]. After a public presentation in Fukushima prefecture, the Cabinet Council dissolved the committee.

The Interim and Final Reports (English versions available from www.cas.go.jp/jp/seisaku/icanps/eng/) are precise records of the investigation into the facts of the accident. The two-volume reports amount to about 1,500 pages of text and figures and are about 8 cm thick. It will be extremely difficult for the general public to read through the pages to understand what the investigation clarified. The facts in the reports are like potatoes and radishes dug from the ground and piled at the storefront of a grocery store. Potatoes and radishes need to be cooked to taste good. In addition, we will not enjoy their flavors unless we consume them. We cannot understand this situation if we are just given the raw material, and unless we can understand the facts, we will never learn anything from this accident, which caused great sacrifices.

In June of 2011, when the committee started, I made an announcement that, as the chair of the committee, my policy was to explain what happened in the course of the Fukushima NPP accident, and why it happened, so everyone could acquire that information as knowledge. In other words, I set the goal to answer the questions of the people of Japan and of the world. The committee, however, primarily due to time constraints and limitations of a temporary organization, could not fully meet that goal to my expectations. I felt that leaving the work as such would not be fulfilling the roles I had assigned to myself at the beginning of the project.

Thus, the authors gathered to make available this book, which plainly explains what we can learn from this catastrophic accident based on what the committee published, in addition to the authors' opinions, which were not in the government reports. The book is targeted toward members of the general public who may not have knowledge in nuclcar power generation or radiology.

Before publishing this book, in December of 2012 Hatamura and Fuchigami published a book titled *What Happened in Fukushima Nuclear Power Plant?: Breakdown of the Government Investigation Committee Report* (in Japanese, coauthored with Naoto Kasahara, Nikkan Kogyo Shimbun Ltd) [3]. Then in April 2013, Hatamura and Fuchigami joined with Seiji Abe to publish *Why Did the Fukushima Nuclear Power Plant Accident Happen?: Breaking Down the Core of the Government Investigation Committee* (in Japanese, Kodansha) [4].

These two books written in Japanese have not been translated into English. The authors felt an obligation to make the knowledge and lessons learned through their investigation available to those worldwide who are interested in the Fukushima NPP accident.

Hatamura, Fuchigami, and Abe, who served as members and technical consultants in the government investigation team, with Kasahara, an expert in nuclear engineering, gathered to edit this English version. This book extracts important contents from the two aforementioned Japanese books, rearranges the significant information coherently, reworks the text into plainer language, and with additional analyses, further clarifies the overall picture of the Fukushima NPP accident. Kenji Iino, a former GE Nuclear engineer, translated our Japanese manuscripts.

The book begins with Chapter 1, which outlines the accident and the operation of the nuclear power industry, including how the government and TEPCO prepared for nuclear safety and disaster prevention. Chapter 2 explains the accident progress inside the plant; Chapter 3 shows where and how the central and local governments made mistakes; and, Chapter 4 outlines problems with TEPCO's response. Chapter 5 discusses the events outside the plant, including how radiation was released and how people evacuated the area. The chapter also discusses how to tackle the problem of decontamination. Chapter 6 reviews the accident from the viewpoint of the "Study of Failure" and reveals my opinions after serving as the committee chair, and finally, Chapter 7 concludes with suggestions on how we should approach (or depart from) nuclear power.

Chapters 1, 3, and 4 were primarily written by Abe; Chapter 2 was written by Fuchigami and Kasahara; and Chapters 5–7 were mainly produced by Hatamura. Position titles and affiliations of people mentioned in the book are all from the time of the accident.

It is the authors' hope that this book will serve to provide the information, clarified as of today, of what happened in Fukushima NPP and how the events progressed. It is intended for those who desire to learn the background and causes of the accident so the information can offer a foundation for those who are at the turning point of whether to push for nuclear power generation or to abandon it. We also hope that the book will provide the lessons we learned from the Fukushima NPP accident to those that are working to reduce accidents and disasters anywhere in the world. It will be our great pleasure if this book can reduce at least some of the anxiety in those who are still not allowed into their homes.

August, 2014
Representing the authors,
Yotaro Hatamura

References

[1] Investigation committee on the accident at the Fukushima Nuclear Power Stations of Tokyo Electric Power Company, Interim report. http://www.cas.go.jp/jp/seisaku/icanps/eng/interim-report.html, 2011 (accessed 20.08.13).
[2] Investigation committee on the accident at the Fukushima Nuclear Power Stations of Tokyo Electric Power Company, Final report. http://www.cas.go.jp/jp/seisaku/icanps/eng/final-report.html, 2012 (accessed 20.08.13).
[3] What Happened in Fukushima Nuclear Power Plant?: Breakdown of the Government Investigation Committee Report (in Japanese), M. Fuchigami, N. Kasahara, Y. Hatamura, and Nikkan Kogyo Shimbun Ltd., December, 2012.
[4] Why did the Fukushima Nuclear Power Plant Accident Happen?: Breaking Down the Core of the Government Investigation Committee (in Japanese), Y. Hatamura, S. Abe, M. Fuchigami, and Kodansha, April, 2013.

The Fukushima-1 nuclear power plant accident

1

S. Abe
Kansai University, Osaka, Japan

1.1 Introduction

This chapter provides background information about nuclear power generation in Japan, including how the Japanese nuclear power industry is structured and the position of the owner of the plant where the accident occurred. The accident was triggered by an earthquake and tsunami that crippled the plant. This chapter also covers the earthquake and tsunami data as background for understanding the accident. Although no direct fatalities were caused by radiation from the plant accident, there were related deaths. Some, however, are hard to categorize into those caused by natural disaster or those that were a consequence of forced evacuation from radiation release by the power plant. The total fatality counts are provided.

The specific troubled plant is outlined here, as well as how Japan was prepared, or at least how it thought it was prepared, for disasters of this kind.

1.2 Energy production in Japan

1.2.1 The energy situation in Japan

Japan consists of 6852 small to large islands with an area of 378,000 km^2. The four main islands are Honshu, Hokkaido, Shikoku, and Kyushu. About 70% of the lands are forests and rivers, and the inhabitable area is only about 30% of the land. As of July 2013, the population was just over 127 million.

Located in the warm region of East Asia, Japan enjoys variations of the four seasons. The temperature in the capital area of Tokyo reaches an average of 27.4 °C in August, and in January it is as low as 6.1 °C (1981–2010). Electricity consumption in Japan thus reaches its peak from July to September followed by a second peak from December to January. The month of lowest electricity consumption is October, when the load is about 60% of the peak value.

The private sector of Japan is responsible for the electricity business. Deregulation in the 1990s invited independent power producers (IPPs) that sell bulk electricity to power companies and some specific corporations that sell electricity to large commercial entities; however, the primary electricity supply is managed by 10 electric power companies, each monopolizing its assigned area (Figure 1.1). These 10 companies

The 2011 Fukushima Nuclear Power Plant Accident. http://dx.doi.org/10.1016/B978-0-08-100118-9.00001-2

Figure 1.1 Japan's 10 electric power companies and their assigned regions.

sold about 860 TWh (tera is 10 to the 12th power) of electricity, about 40% of which was to general consumers and the remaining 60% to industry.

Japan relies on importing most of its fossil fuel from foreign countries. In order to secure energy sources and reduce CO_2 exhaust that causes the greenhouse effect, Japan has been actively introducing nuclear power generation since the 1970s. As a result, Japan is now the third-largest nuclear-powered country, next to the United States and France, with about 30% of its electrical power generated by over 50 nuclear power plants (NPPs) (before the Fukushima accident). Figure 1.2 shows those NPPs that are referenced in this book and a fossil fuel plant, Haramachi, that suffered damage. Tokyo Electric Power Company (TEPCO) owned Fukushima-1 (Fukushima Daiichi) NPP, where the accident occurred.

1.2.2 The Fukushima-1 accident: an unprecedented nuclear power accident

On March 11, 2011, a series of tsunami caused by the Tohoku Area Pacific Offshore Earthquake destroyed all the off-site and almost all the internal power sources in Fukushima-1. These losses led to a failure to cool the reactors and the spent fuel storage pools and eventually to a Level 7 major accident on the International Nuclear and Radiological Event Scale (INES). Fukushima-2 (Fukushima-Daini) NPP was also damaged and experienced a Level 3 serious incident.

The Three Mile Island accident (Level 5) in 1979 and Chernobyl accident (Level 7) in 1986 were both single-reactor accidents; in contrast, Fukushima-1 involved, for the first time worldwide, damage to three reactors at the same time. The accident forced a large number of local residents to evacuate the area. The number of evacuees, including those from the earthquake and tsunami, reached, at its peak, over 90,000 to other prefectures and over 60,000 within the Fukushima prefecture.

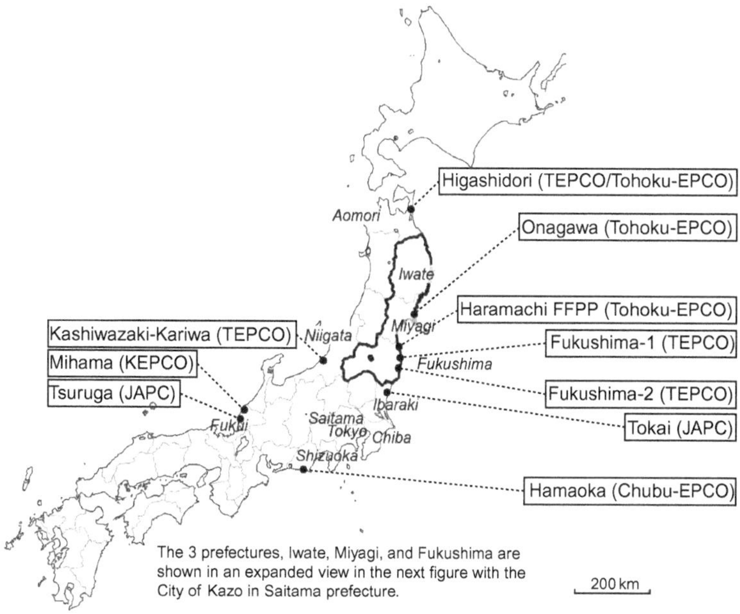

Figure 1.2 Power plants discussed in this book. Prefectures are shown in italic, while power plants (with operator in parentheses) are shown within boxes. Haramachi FFPP is a fossil fuel power plant. Three prefectures, Iwate, Miyagi, and Fukushima, are shown in an expanded view in Figure 1.3 with the city of Kazo in Saitama prefecture.

The number of disaster-related deaths in Fukushima prefecture also shows the severity of the Fukushima-1 NPP accident. The term *disaster-related death* has a distinct definition in that the death was caused not directly by an earthquake or tsunami but by a later indirect event, and the bereaved family has received condolence money in accordance with the "Regulation about condolence money payment for disaster-related deaths" [1]. After the earthquake, the Reconstruction Agency announced that the number of Great East Japan Earthquake-related deaths was 3,089 as of March 31, 2014. The number included 1,704 in Fukushima, followed by 889 in Miyagi, 441 in Iwate, 41 in Ibaraki, and so on. Table 1.1 contrasts the number of deaths and disappearances caused directly by the Great East Japan Earthquake and the disaster-related deaths in each prefecture. The disaster-related deaths in Fukushima prefecture clearly stand out compared with either Miyagi or Iwate.

Table 1.2 shows the breakdown of the disaster-related deaths in Fukushima prefecture into smaller districts. The municipal regions adjacent to Fukushima-1 (Minamisoma, Namie, Tomioka, Futaba, Naraha, and Okuma), show large numbers of disaster-related deaths compared with the direct fatalities. These tables show how the NPP accident caused large sacrifices to the people in the region.

Table 1.1 Victims of the Great East Japan Earthquake

	Deaths	Disappeared	Disaster-related deaths
Fukushima	1,609	207	1,704
Miyagi	9,538	1,269	889
Iwate	4,673	1,132	441
Ibaraki	24	1	41
Others	43	3	14

Note: Deaths and disappeared counts are as of July 10, 2014 and the number of disaster-related deaths are as of March 31, 2014. Miyagi, Iwate, and Fukushima suffered large numbers of victims.
Adapted from Source: Refs. [16,17].

Table 1.2 Fatality counts in local districts of Fukushima prefecture

	Deaths				
	Direct	Disaster related deaths	Reported: no body found but reported dead	Total deaths	Population
Caution zone districts within 20 km of Fukushima-1					
Minamisoma	525	457	111	1,098	70,878
Namie	149	329	33	511	20,905
Tomioka	18	250	6	274	16,001
Futaba	17	107	3	127	6,932
Okuma	11	103	-	114	11,515
Naraha	11	100	2	113	7,700
Kawauchi	-	72	-	72	2,820
Districts within 30 km of Fukushima-1					
Iwaki	293	125	37	455	3,42,249
Iitate	1	42	-	43	6,209
Hirono	2	39	-	41	5,418
Katsurao	-	28	1	27	1,531
Tamura	-	9	-	9	40,422
Total	1,027	1,661	193	2,884	5,32,580

Note: The total count does not match that of Table 1.1 due to counting on different dates.
Adapted from Source: Refs. [18,19].

As we mentioned earlier, Japan consists of four main islands and a number of smaller ones around them. Its land is divided into 47 prefectures. Forty-six of them, excluding Okinawa, are shown with gray borders in Figure 1.2. Each prefecture is further divided into regional districts. Figure 1.3 shows all the regional districts referenced throughout this book.

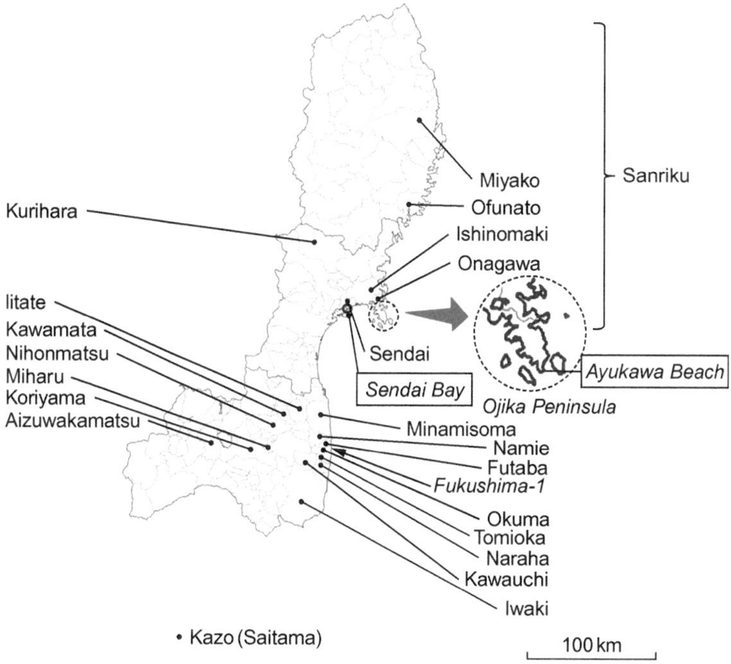

Figure 1.3 Districts discussed in this book. The city of Kazo is shown in the geographically correct location.

1.3 The Fukushima-1 nuclear power plant

1.3.1 TEPCO and nuclear power generation

TEPCO is the largest electric power company in Japan supplying electricity to the Tokyo metropolitan area and surrounding regions. Its electric power sales amounted to 293 TWh (Japan Fiscal Year (JFY) 2010; JFY 2010 covers April 1, 2010 to March 31, 2011), which is about one-third of the total amount in Japan and is roughly equivalent to the entire consumption of Italy.

In JFY 2010, the licensed power generation facilities of TEPCO included fossil fuel of 59.5%, nuclear of 26.6%, hydraulic of 13.8%, and others at 0.1%. During the same fiscal year, the overall ratios for the 10 utility companies were fossil fuel 60.2%, nuclear 22.4%, hydraulic 17.1%, and others 0.3%. Thus, among the 10 companies, TEPCO had a slightly lower percentage of hydraulic and a slightly higher share of nuclear. The actual ratio of reliance on nuclear power before the accident (excluding the licensed but stopped plants) was 28% with TEPCO. The breakdown of actual power generation during that fiscal year with the 10 electric companies was fossil fuel 59.0%, nuclear 33.0%, hydraulic 7.7%, and others 0.3%.

As Table 1.3 shows, at the time of the accident in March 2011, TEPCO had three power plants in operation: Fukushima-1 and Fukushima-2 in Fukushima prefecture

Table 1.3 TEPCO NPPs in operation

Plant name	Location	Start-up of first unit	Number of units	Licensed power
Fukushima-1	Futaba, Fukushima	1971	6 (BWR)	4696 MW
Fukushima-2	Futaba, Fukushima	1982	4 (BWR)	4400 MW
Kashiwazaki-Kariwa	Kariwa district, Kashiwazaki-shi, Niigata	1985	5 (BWR) 2 (ABWR)	8212 MW

This table is based on the web data from NRA [20–22].
BWR: boiling water reactor; ABWR: advanced boiling water reactor.

and Kashiwazaki-Kariwa in Niigata prefecture. The company also had begun construction in January 2011 of Higashidori NPP (two units of advanced boiling water reactors (ABWRs) licensed to produce 2770 MW) in Shimokita district in Aomori prefecture.

1.3.2 Overview of Fukushima-1 nuclear reactors

Fukushima-1 NPP stands on land that covers parts of Okuma and Futaba of Futaba district in Fukushima prefecture. It is the oldest among the three NPPs currently in operation by TEPCO. Fukushima-1 had six units of licensed boiling water reactors (BWRs), as Table 1.4 shows. Construction on the oldest, Unit 1, began in 1967, and commercial operation began in 1971. The newest, Unit 6, started commercial operation in 1979.

Table 1.4 Nuclear reactors at Fukushima-1

	Unit 1	Unit 2	Unit 3	Unit 4	Unit 5	Unit 6
Licensed	1968	1967	1970	1972	1971	1972
Start of construction	1967	1969	1970	1972	1971	1973
Start of commercial operation	1971	1974	1976	1978	1978	1979
Power (MW)	460	784	784	784	784	1,100
Containment vessel	Mark I	Mark I	Mark I	Mark I	Mark I	Mark II
Japan made (%)	56	53	91	91	93	63
Primary manufacturer	GE	GE Toshiba	Toshiba	Hitachi	Toshiba	GE Toshiba
Number of fuel bundles	400	548	548	548	548	764

Source: Ref. [23].

After this accident, Units 1 to 4 were retired on April 19, 2012. Thus, as of now, Fukushima-1 NPP has only two licensed units – Units 5 and 6.

Figure 1.4 shows the layout of Fukushima-1 NPP. The plant faces the Pacific Ocean to the east, and of the 6 units, 1 to 4 are in Okuma, and 5 and 6 are in Futaba. Reactors 1 to 4 stood from north to south in that order and 5 and 6 from south to north. In March 2011, these six reactors had a total licensed power capacity of 4696 MW, which was ranked in third place among the 17 NPPs in Japan.

Each unit consists of a reactor building, turbine building, control building, service building, nuclear water management building, and so on. Some buildings were shared among adjacent units. The entire plant covered a wide area of about 3,500,000 m^2 shaped in a semi-ellipse with its major axis along the coast.

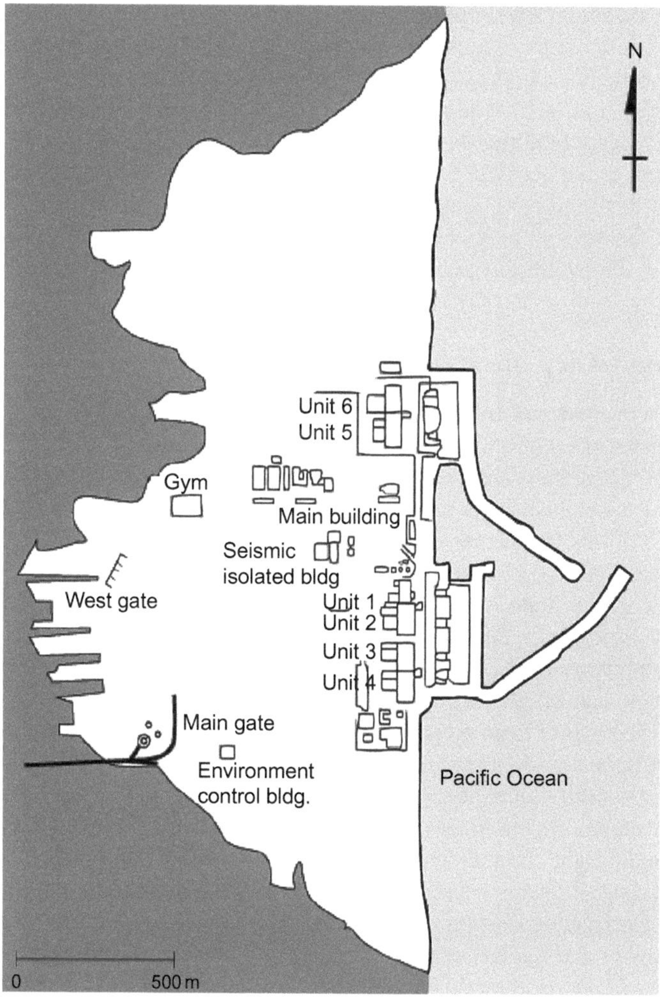

Figure 1.4 Layout of Fukushima-1.
Source: TEPCO.

1.3.3 Operation of the NPP

The number of TEPCO employees working at Fukushima-1 at the time of the accident was about 1,100. Other workers at the plant included maintenance, fire protection, security guards, and other contractors summing the workforce count to about 2,000. The operation of Fukushima-1 relied on a large number of subcontracted workers.

During normal operation, two unit managers and three associate managers reported to the plant manager. The organization had departments for administration, disaster prevention and safety, public relations, quality and safety, engineering, operation management 1 and 2, and maintenance 1 and 2.

Operators on duty controlled the reactor facilities. They were all TEPCO employees assigned to unit pairs of 1 and 2, 3 and 4, and 5 and 6, and they reported to operation manager 1 and operation manager 2.

On-duty operators were assigned to groups, each with 11 employees: a leader, a subleader, two chief operators, an assistant chief operator, two primary facility operators, and four auxiliary facility operators. The groups were on shift to operate and manage the reactor facilities 24 h a day.

At the time of the accident on March 11, units 4, 5, and 6 were in periodic maintenance, and the number of people on-site then was larger than the usual head count. About 6,400 workers were on-site, including 750 TEPCO employees; about 2,400 of them were in the radiation-controlled area.

1.3.4 Emergency operations

TEPCO, as a nuclear power utility company, was required to follow the basic law of nuclear disaster prevention, i.e. the Act on Special Measures concerning Nuclear Emergency Preparedness (Nuclear Emergency Preparedness Act) [2], which required Fukushima-1 to set its disaster prevention plans. When an unusual event as described in Article 10 of this law is reported, a Class 1 Nuclear Emergency State is declared.

In the case of Article 15 or declaration of the state of Nuclear Emergency, a Class 2 Nuclear Emergency State is announced. These announcements trigger the setting up of a Nuclear Emergency Response Headquarters (NERHQ) for removal of the accident cause, preventing the spread of nuclear disaster, and taking other necessary actions quickly and effectively.

The NERHQ was to form groups for information, reporting, announcements, engineering, safety, restoration, power generation, procurement, health, medical, administration, and security/guidance. Each group was to carry out its role to establish a disaster-prevention system in the case of a nuclear accident. The head of the NERHQ was to maintain tight communication with the staff dispatched to the off-site center (Nuclear Disaster Prevention Staff) to take the actions requested by the Joint Council for Nuclear Emergency Response and to report status and suggestions to the council.

Operations of the nuclear reactor facilities, at the time of emergency, were the responsibility of the operator on duty in the same manner as when the plant was under normal operations.

1.4 The Tohoku Area Pacific Offshore Earthquake and tsunami

1.4.1 The Tohoku Area Pacific Earthquake

At 14:46 on March 11, 2011, a great earthquake hit an area offshore from Sanriku coastline (about 130 km east-southeast of Ojika peninsula) with a 24-km-deep seismic center. The earthquake shook Kurihara in Miyagi prefecture with magnitude 7, and other areas along the Pacific coast of Tohoku area suffered strong tremors of magnitude 6.

The Japan Meteorological Agency named the earthquake "Tohoku Area Pacific Offshore Earthquake," and the cabinet, on April 1, 2011, approved the name of "Great East Japan Earthquake" to the disasters caused by this earthquake.

1.4.2 The Tohuku Area tsunami

Tsunami waves induced by the Tohoku Area Pacific Offshore Earthquake hit the entire eastern side of Japan from Hokkaido to Okinawa, and the area from Tohoku to Chiba prefecture was faced with huge waves. The Japan Meteorological Agency reported an 8 m plus tsunami in Miyako and Ofunato of Iwate prefecture, 9.3 m or higher in Soma of Fukushima prefecture, and 8.6 plus in Ayukawa of Ishinomaki, Miyagi prefecture.

The Geospatial Information Authority of Japan reported that the tsunami flooded a total land area of 561 km^2 covering 62 local districts in the six prefectures of Aomori, Iwate, Miyagi, Fukushima, Ibaraki, and Chiba. The two prefectures of Miyagi (327 km^2) and Fukushima (112 km^2) suffered huge areal flooding [3]. The earthquake and tsunami took the lives of many people. According to the National Police Agency announcement on July 10, 2014; 15,887 people died in 12 prefectures, and 2656 disappeared in 6 prefectures, mostly in Iwate, Miyagi, and Fukushima.

The number of deaths in the 1995 Great Hanshin-Awaji Earthquake was 6434, and the final number of disappearances was 3 (finalized by the Fire and Disaster Management Agency on May 19, 2006). This earthquake hit the city of Kobe and the surrounding metropolitan area. Most of the bodies were found under the wreckage of buildings, and the number of disappearances was small. The large number of disappearances of 2600 or more with the Great East Japan Earthquake is an indication of the large number of tsunami victims.

As an aside, nuclear power generation requires a heat-sink (the final place to release heat), and all NPPs in Japan are built next to the ocean. In March 2011, five plants were in operation along the Pacific coast of the Tohoku area: Higashidori, Onagawa (Tohoku-EPCO), Fukushima-1, Fukushima-2 (TEPCO), and Tokai-2 (The Japan Atomic Power Company, JAPC). The March 11 tsunami affected not just Fukushima-1 but also the other four NPPs. Here, we will describe what happened with these other four plants.

First at Higashidori, tsunami waves arrived but they did not go over the top of the cliff (T.P.* +2.6 m), and only a small amount of damage was reported. The tsunami waves that hit Onagawa reached about 13 m, a level higher than the design height of 9.1 m, but the primary facilities of Onagawa stand, for safety precautions, on land at 13.8 m (measured after the earthquake caused subsidence), and thus that plant did not suffer serious damage.

At Fukushima-2 located about 10 km south of Fukushima-1, the ocean-side facilities with seawater pumps were submerged underwater; however, the plant was built on land on the mountainside up a slope that was 2 m higher than Fukushima-1. The tsunami did not reach the reactor building, and the plant only suffered INES Level 3 events.

Last at Tokai-2, the tsunami came up to a height of 6.3 m; however, it did not reach the primary buildings, and the plant suffered no severe damage. As we will discuss later, Fukushima-1, with insufficient preparation against tsunamis and severe accidents, faced a series of severe accidents.

1.4.3 Status of the nuclear reactors before the earthquake

As mentioned earlier, Fukushima-1 had six nuclear reactors. We will review the status of each unit immediately before the earthquake.

- Unit 1 was at steady-state operation at rated electrical power. This operation mode keeps the electricity generation at the rated electrical output available throughout the year. The fuel pool, next to the reactor for storing spent fuel, was filled with water at 25 °C.
- Units 2 and 3 were both at steady-state operation at rated thermal power. This mode keeps the reactor thermal output at the maximum value allowed according to the reactor licensing standards. Both fuel pools were filled with water with temperature values of 26 °C for Unit 2 and 25 °C for Unit 3.
- Unit 4 had been in annual maintenance since November 30, 2010. Annual maintenance is conducted roughly once a year in accordance with the Electricity Business Act to verify that facilities and equipment are in good condition while meeting the required functionalities and improving their reliability. For this maintenance, all fuel bundles were removed from the reactor pressure vessel (RPV) and stored in the spent fuel pool. The water level of the pool was full with the temperature at 27 °C.
- Unit 5 had been in annual inspection since January 3, 2011. The fuel bundles, however, were in the reactor with all control rods fully inserted for a pressurized leakage test. The fuel pool was full of water at 24 °C.
- Unit 6 had also been in annual inspection since August 14, 2010. The reactor was at cold shutdown, fully loaded with fuel and control rods all fully inserted. The fuel pool was full with water at 25 °C.

1.4.4 Seismic movement at Fukushima-1 NPP

The maximum intensity of Tohoku Area Pacific Offshore Earthquake measured in Okuma and Futaba districts where Fukushima-1 stood was "magnitude 6 strong."

*T.P.: Tokyo Peil. The average sea level of Tokyo Bay, which sets the standard sea level of Japan.

After the first movement, a number of aftershocks at "magnitude 5 weak" continued to shake the area for days. "Magnitude 6 strong" is the second-highest intensity on the earthquake magnitude scale. The fact that the two districts measured this level of intensity indicates that an extremely strong seismic movement took place in this area.

The Japan Meteorological Agency lists the following as symptoms of "magnitude 6 strong":

- Have to crawl to move. May get thrown away.
- Most unsecured furniture moves and many items fall down.
- Many wooden buildings tilt or fall over.
- Cracks in ground, large-scale landslides, or collapse of hills may take place.

Fukushima-1 had 53 seismographs on site – on the ground and in the reactor and turbine buildings for every unit – to monitor seismic movements. Table 1.5 shows the maximum acceleration recorded at the base of the reactor building for each unit. The table shows the recorded maximum acceleration on units 2, 3, and 5 in the east–west direction exceeded the short-period spectral acceleration (Ss) of the maximum expected earthquake where underlined.

1.4.5 The tsunami at Fukushima-1 NPP

The first tsunami wave hit Fukushima-1 at around 15:27 on March 11. The second wave hit at about 15:35, and further waves continued thereafter. The wave that devastated Fukushima-1 was the second wave. The ocean-side area of Fukushima-1 and almost all its primary building area went underwater. The damage progressed in the following manner.

Table 1.5 Recorded acceleration and maximum response acceleration with short-period spectral acceleration (Ss) of the maximum considered earthquake

Reactor	Recorded maximum acceleration (Gal)			Maximum response acceleration with short-period spectral acceleration (Ss) of maximum considered earthquake (Gal)		
	South–north	East–west	Vertical	South–north	East–west	Vertical
Unit 1	460	447	258	487	489	412
Unit 2	348	550	302	441	438	420
Unit 3	322	507	231	449	441	429
Unit 4	281	319	200	447	445	422
Unit 5	311	548	256	452	452	427
Unit 6	298	444	244	445	448	415

Source: Ref. [24].

The inundation height in the primary building area of Units 1 to 4 was O.P. + 11.5 to +15.5 m. (O.P. stands for Onahama Peil, the construction base level at Onahama Bay in Iwaki, Fukushima prefecture, located about 50 km south from Fukushima-1.) The land level of the area is about O.P. +10 m; thus, the inundation depth was 1.5 m where shallow and at some places reached as deep as 5.5 m. The inundation height even reached O.P. +16–17 m in some spots southeast of the primary building area. This area suffered the deepest submergence in Fukushima-1.

Units 5 and 6 were built on a different block from that of Units 1 to 4. Their primary building area had an inundation height of about 13–14.5 m. The land of this area is O.P. +13 m; thus, the inundation depth was 1.5 m or less. Units 5 and 6 succeeded in cold shutdown even though they were affected by a tsunami of the same level as Units 1 to 4. One of the reasons for this was because the primary buildings of Units 5 and 6 were built on relatively high land.

The safety of nuclear power facilities depends on the three basic principles of "stopping," "cooling," and "containing." Fukushima-1 was attacked by Tohoku Area Pacific Offshore Earthquake and the great tsunami waves that followed the quake. The plant succeeded in "stopping" by quickly inserting the control rods (SCRAM) into the core immediately after the earthquake's arrival. The plant, however, failed in its "cooling" function due to damage from the earthquake and water damage to the electric power sources caused by the tsunami. Three reactors were damaged, and radioactive materials were released to the surrounding environment. In other words, the plant failed in attaining the principles of "cooling" and "containing." Chapter 2 will discuss the details of these processes after the tsunami attack; that is, how the situation progressed in Fukushima-1 and how this led to severe accidents.

1.5 Nuclear power safety and disaster prevention systems in Japan

1.5.1 Nuclear safety laws and regulations

When accidents or problems occur in nuclear power facilities, radioactivity spread to their surroundings cause serious damage to people and the environment. Ensuring safety, therefore, is subject to strict regulations. The following sections describe the Japanese legal system regulations about the safety of nuclear power generation.

First, at the very top is the Atomic Energy Basic Act [4], which took effect in 1956. This act is literally the basic laws of using nuclear power by setting the fundamentals of research, development, and utilization of nuclear power. A number of regulations followed this act: the Act on the Regulation of Nuclear Source Material, Nuclear Fuel Material and Reactors [5], effective 1957; the Act on Prevention of Radiation Disease Due to Radioisotopes, etc. [6], effective 1958; and the Designated Radioactive Waste Final Disposal Act [7], effective 2000. Also, the Electricity Business Act [8], passed in

1964, is the basic law relating to the control of electricity businesses electricity businesses, but it also covers the basics of regulating nuclear power facilities from the standpoint of machines for electricity generation.

These series of acts were followed by ordinances like the Order for Enforcement of the Act on the Regulation of Nuclear Source Material, Nuclear Fuel Material and Reactors [9] and the Order for Enforcement of the Act on Prevention of Radiation Disease Due to Radioisotopes, etc. [10] and rules, for example Rules on Building and Operating Practical Electricity Generating Nuclear Reactors [11] or Rules on Business of Storing Spent Fuel [12].

The Nuclear Safety Commission (NSC), which was closed in September 2012, had set guidelines that the regulating body, the Nuclear and Industrial Safety Agency (NISA), used for safety reviews, and those guidelines were also used for safety control.

1.5.2 Administration of nuclear safety

In Japan, the Minister of Economy, Trade and Industry (METI) governs the power-generating reactors in the industry, and the Minister of Education, Culture, Sports, Science and Technology (MEXT) controls research, development, related use, and radiological protection and preparation. Within this structure, NISA's position under the Agency for Natural Resources and Energy of METI was as a special organization to regulate the safety of nuclear reactor facilities for power generation.

When the central administration was reformed in 2001, NISA was established to take over the administration of safety with high-pressure gas, urban gas, liquefied petroleum gas (LPG), explosives, and mines, which had previously been the responsibility of the Nuclear Safety Bureau of Science and Technology Agency or the Environmental Protection and Industrial Location Bureau of METI. NISA, at the same time, also took over safety administration of electrical facilities, urban gas, and heat supply from the Resources and Energy Agency. The organization, thus, regulated safety of not only nuclear power but also gas, mines, explosives, and other energy-related industries. Furthermore, in addition to safety regulation, NISA was expected to assume the central role in administering the NERHQ.

Actions during the Fukushima NPP accident, however, revealed the organizational limitation of NISA, which failed to perform the assigned roles. In addition, questions were raised about its regulatory activities before the accident. The organization was closed on September 19, 2012, and its operations were transferred to the new Nuclear Regulation Authority (NRA; established September 19, 2012), an external bureau of the Ministry of the Environment.

In addition to NISA, an incorporated administrative agency, the Japan Nuclear Energy Safety Organization (JNES), had been established in 2003 as a public organization for safety regulation of nuclear energy. JNES took roles as an organization to provide technical assistance to NISA (e.g., conducted inspections of nuclear power facilities jointly with NISA and provided technical assistance in organizing regulations on safety inspection and control of nuclear facilities). As of April 2012, JNES had about 423 officers and employees (i.e., about the same number of those at NISA when it was closed). JNES survived the 2012 reformation of nuclear safety regulatory organizations but later on March 1, 2014, was merged into NRA.

Although it does not conduct direct regulation on business entities, the NSC Safety Commission was another organization that played a role in the safety of nuclear power. The NSC was established in 1978 as an independent organization to enhance organizational nuclear safety by separating this function from the former Atomic Energy Commission. The establishment was based on such laws as the Atomic Energy Basic Act and the Act for Establishment of the Japan Atomic Energy Commission and the Nuclear Safety Commission [13].

Nuclear safety regulation, as we mentioned earlier, was carried out by NISA, MEXT, and other administrative organizations. Among them, the NSC held an independent neutral position to plan, discuss, and determine the basics of safety regulation by the government, and at the same time it conducted the second review (double-checking) for Applications for License to Build a Nuclear Reactor and carried out regulatory inspections. In other words, this organization supervised, audited, and instructed both utility companies and administrative organizations. The NSC, thus, had the power to issue adjuration to related administration offices in the name of the prime minister.

The functional limitations of the NSC, similar to NISA, were exposed by the Fukushima NPP accident, and the commission was shut down on September 19, 2012, with their operations transferred to the NRA.

1.5.3 Organizations in charge of nuclear safety and regulation

Safety regulations of commercial power generation are categorized in two types in countries around the world; one type uses private corporations to regulate the industry, and in the other, nationally owned organizations perform the task. In the latter case, the government directly regulates power generation, whereas in the former case, government involvement is indirect through regulation of public utilities. In the case of Japan, as of 2014, 10 electric companies carry out the overall business from power generation to sales to the consumer. Each of these 10 companies has its own exclusive territory.

Of these 10 electric companies, 9, excluding Okinawa Electric Power, contribute to the commercial nuclear power generation business. As declared in the 2006 IAEA Fundamental Safety Principles, Principle 1 (Responsibility for Safety), the first in line of responsibility for safety of nuclear plants is the electric company. This principle holds for electric companies in Japan as well. The business of nuclear power generation is subject to regulations of public business as utility companies, and in addition, they also have to follow strict government safety regulations to secure a higher level of safety.

At the time of the Fukushima-1 accident on March 11, 2011, the government organization in charge of regulating the nuclear power generation business was NISA. NISA, following the Act on the Regulation of Nuclear Source Material, Nuclear Fuel Material and Reactors and the Atomic Energy Basic Act, conducted safety regulation of nuclear power facilities for the three stages of their lives: (1) during design and construction, (2) during operation, and (3) during shutdown decommissioning. Also, another organization, the NSC, conducted regulation separate from NISA. NSC's primary roles were:

1. In the licensing stage, conduct the second review after the first by the administration agent (i.e., double-checking).
2. Supervise regulation by the administration agent.
3. Respond to nuclear power emergency situations.

The Fukushima accident showed that the regulating organizations lacked adequate levels of technical expertise in their regulating abilities. NISA and NSC were judged to have performed poorly, and as we described earlier, they were abolished in September 2012, and a new regulating organization, the Nuclear Regulation Authority (NRA) and its administration office, the Nuclear Regulation Agency, were formed.

1.5.4 Overview of the legal system for nuclear disaster management

The legal system in Japan regarding nuclear disaster management has expanded from the foundation of the Basic Act on Disaster Control Measures [14], set in 1961, and the Act on Special Measures concerning Nuclear Emergency Preparedness (the Nuclear Emergency Preparedness Act) in 1999.

The Basic Act on Disaster Control Measures assigns the responsibility as follows: to the Central Disaster Management Council, headed by the prime minister, for setting the basic plans in disaster control; to each Prefectural Disaster Management Council for setting the prefectural plans in disaster control; and to each smaller district for setting regional plans. Actions to take in the case of nuclear disasters are part of these plans. The basic, prefectural, and regional plans have sections on Countermeasures against Nuclear Disasters among other sections about general disasters, earthquakes, and accidents. These plans lay out the basics of countering nuclear disasters by preventing their occurrence and expansion and by necessary measures to take in the case of nuclear disasters to plan recovery.

These facts, however, do not mean that the Central Disaster Management Council will take action on nuclear disasters. The Basic Act on Disaster Control Measures calls for each prefecture to respond to nuclear disasters. The Central Disaster Management Council is not an organization to carry out the actual disaster countermeasures. The council would only take part in evaluating the construction of new NPPs.

Fukushima prefecture, thus, definitely had an important responsibility in taking actions to prevent this nuclear disaster; however, as we will discuss in Chapter 3, the prefecture's countermeasures were not effective enough. Reflecting on its performance, the prefecture, after the accident, started to make significant revisions in their Local Disaster Management Plans and Nuclear Disaster Countermeasures. For the first step, the prefecture announced its new local disaster management plans in November 2012. The revised plans, on the current status of Fukushima-1, clarified that the disaster management is for nuclear reactor facilities that have been determined to be decommissioned and those that are currently shut down. Learning from the fact that insufficient preparations were in place against complex disasters, the plans define a Nuclear Group to be located in the prefectural headquarters administration office and include new policies to centralize plant status and monitoring functions. Fukushima prefecture also announced plans for further review of the disaster management plans, including evacuation criteria and identifying critical locations.

Next, we will review the Nuclear Emergency Preparedness Act. Japan, for a long time, in the case of a nuclear disaster, took actions based on the Basic Act on Disaster Control Measures. In 1999, in Tokai district of Ibaraki prefecture, a critical accident took place in JCO's nuclear fuel processing facilities, and thus the Nuclear Emergency Preparedness Act was born. Since then, countermeasures against nuclear disasters have been based on this regulation, and the system of nuclear disaster management has dramatically changed.

The Nuclear Emergency Preparedness Act consists of 40 articles and addendums that set the basics of countering nuclear disasters, such as obligations of nuclear power utility companies for preventing nuclear disasters, issuance of declaration of nuclear emergency situations, setting up the NERHQ, and executing emergency response measures. The purpose is to enhance the preparedness against nuclear disasters and thereby protect the lives, bodies, and properties of citizens from them.

The council of governmental bodies involved with risk management for nuclear disasters prepared a *Nuclear Emergency Response Manual* [15]. This manual summarizes the actual points necessary for related governmental bodies to cooperate in conducting a consistent set of disaster management actions in the case of a nuclear disaster, based on the Nuclear Emergency Preparedness Act and Disaster Management Plans, and Nuclear Disaster Countermeasures.

1.5.5 Structure of the Nuclear Emergency Preparedness Act

As mentioned earlier, the Nuclear Emergency Preparedness Act sets the basics of countermeasures to take if a nuclear disaster takes place in Japan. The law underwent a major revision in June 2012, following the Fukushima NPP accident. Here we will discuss the structure of the law before the revision.

The Nuclear Emergency Preparedness Act listed the following three bodies as the primary organizations to counter nuclear disasters. First is the Cabinet Office, METI, NISA, and other government organizations. Next were the local governments where nuclear plants were located and, third, nuclear power utility companies that operated the plants and related offices.

Table 1.6 lists the articles describing the emergency actions for these three bodies as defined in the Nuclear Emergency Preparedness Act. As we can see in the table, the governmental bodies assumed important roles, and local governments also had large responsibilities.

The Nuclear Emergency Preparedness Act assigned the core organization of nuclear disaster countermeasures to the NERHQ set by the prime minister. The following section summarizes the events from the disaster breakout to the setting up of the NERHQ.

First, Article 10, Section 1 stated that on detection of a radiation dose above the limit specified by a Cabinet Order (5 μSv per hour) near the border of a nuclear site, a nuclear emergency preparedness manager had to immediately notify the minister,

Table 1.6 Main emergency actions set by the Nuclear Emergency Preparedness Act

Primary agent	Article-section	Actions
Central government, METI	10-1	Receive notification
	10-2	Dispatch expert officials
	15-1	NISA to submit drafts of a public notice and an instruction
	15-2	Prime minister to declare a nuclear emergency situation
	15-3	Prime minister to instruct evacuation or taking shelter
	16	Set up NERHQ within the Cabinet Office
	17-8	Set up local nuclear emergency response headquarters
	20-3	Request assistance from other organizations
	23	Set Joint Council for Nuclear Emergency Response
	26	Execute emergency response measures
	27	Execute measures for restoration from nuclear emergency
Local government	10-1	Receive notification
	10-2	Request dispatch of expert officials
	22	Set up prefectural and district headquarters for disaster control
	23	Set up Joint Council for Nuclear Emergency Response
	26	Execute emergency response measures
	27-1	Execute measures for restoration from nuclear emergency
	28	Instruct evacuation and request disaster relief operation
Nuclear power utility company	10-1	Notify minister, prefectural governors, and district mayors
	25	Execute emergency measures for preventing nuclear disaster expansion
	26	Execute emergency response measures
	27-1	Execute measures for restoration from nuclear emergency
	28	Execute emergency response measures for designated public organizations, and report damages

Source: Ref. [2].

the prefectural governor, the district mayor, and the related neighboring prefectural governors about the detection (Article 10 notification).

Next, Article 15, Section 1 stated that on detection of a radiation dose above 500 μSv per hour in the aforementioned area, or when injecting neutron absorbent could not stop a reactor core, the minister should immediately report necessary information concerning the situation to the prime minister and submit drafts of a public notice and an instruction. Then the prime minister should declare a nuclear emergency situation (Article 15, Section 2) and temporarily set up a NERHQ within the Cabinet Office (Article 16, Section 1).

1.5.6 Nuclear Emergency Preparedness Act guidelines on emergency measures

As we mentioned earlier, Article 16, Section 1 stated that when the prime minister declares a nuclear emergency situation, the cabinet sets up the NERHQ in its office. The prime minister and the competent minister, respectively, would fulfill the duties of the head and the second in command. Other members included ministers of the State, the deputy chief cabinet secretary for crisis management, and heads of designated administrative organizations appointed by the prime minister (Article 17, Sections 1, 4, 6).

Next, the Nuclear Emergency Preparedness Act Article 17, Section 9 is stated to set up Nuclear Emergency Response Local Headquarters (local NERHQ) within the area where emergency response measures would be carried out, to take part in administration of NERHQ responsibility. Article 22 of the Nuclear Emergency Preparedness Act also stated that prefectures and districts that have jurisdiction over areas where emergency response measures would be carried out should also set up prefectural or district emergency response headquarters. In addition, prefectural and district emergency response headquarters should organize a Joint Council for Nuclear Emergency Response for sharing information about the nuclear emergency situation and mutual cooperation in their emergency response measures (Article 23).

The Nuclear Emergency Preparedness Act Article 12, Section 1 further obligates the government to set up a facility (off-site center) to serve as the center for emergency response measures as a central office of gathering information on a nuclear disaster. Ordinance for the enforcement of the Nuclear Emergency Preparedness Act limited the off-site center to within 20 km from the NPP. Fukushima-1 had a joint off-site center shared with Fukushima-2 at about 5 km from Fukushima-1 and about 12 km from Fukushima-2 in Okuma, Futaba district, Fukushima prefecture. The aforementioned local response headquarters (Article 17, Section 9) and Joint Council for Nuclear Emergency Response (Article 23, Section 4) were to be set up at this off-site center.

The preceding discussion was the biggest point of the Nuclear Emergency Preparedness Act (i.e., at a time of nuclear disaster, the central and local governments would closely coordinate, and with the local off-site center in the middle, prefectural and district emergency response headquarters and the Joint Council for Nuclear Emergency Response would be formed to counter the disaster).

The off-site center location was selected from lessons learned from the JCO criticality accident that disaster measures are more effective when conducted close to the actual point where the disaster occurred. The expectations were high for the center in assuming a central role; however, as we will discuss in Chapter 3, it struggled to meet that role.

1.6 Conclusions

The earthquake that hit eastern Japan on March 11 was one of the largest that the world had experienced with a magnitude of 9.0. It was the biggest earthquake ever on record in Japan, and was followed by the tsunami that swept the towns along the Pacific coastal line of northeast Japan. Some ran for their lives and survived the waves. Others, for reasons such as relying on seawalls that were too low and fragile, listening to early warnings that underestimated the height of the tsunami, or being unable to move from the beds they were confined to, lost their lives.

Some took the right precautions and that was why they ran toward higher ground, but for others, it was a total surprise. TEPCO, the owner of a number of huge multiunit NPP sites, was insufficiently prepared for the tsunami that affected Fukushima-1 NPP. Fukushima-1, however, was not the only plant affected. The other plants succeeded in stopping, cooling, and containing radiation within their nuclear reactors. When we study accidents, we usually analyze the failures and point out what went wrong. At the same time, however, we shall take a close look at the contrasting successes and study what went right. This time, Fukushima-1 was the one that failed in cooling the reactor and containing radiation. Why did this particular plant fail and cause a catastrophic nuclear accident?

Japan as a nation, at a glance, seemed well prepared against such natural disasters. This chapter outlined how Japan was prepared. Legislation to control the nuclear industry started as early as in 1956, and a number of regulations have been put in effect since then. In particular, the Nuclear Emergency Preparedness Act that came into effect in 1999 following the JCO criticality accident clearly defined the roles of the central and local government as well as those of the utility companies. NISA and NSC were created as the regulating bodies for the nuclear industry (and others in the case of NISA) to ensure the safety of society. The rest of this book focuses on the questions of "Why did Fukushima-1 fail?" and "What can we do to avoid similar events in the future?"

References

[1] Regulation about condolence money payment for disaster-related deaths, September 18, 1973.

[2] Act on Special Measures concerning Nuclear Emergency Preparedness (Nuclear Emergency Preparedness Act). http://www.japaneselawtranslation.go.jp/law/detail_main?re=2&vm=02&id=106, 1999 (accessed 17.12.13).

[3] Flooded Area by Tsunami (Report Number 5, Rough values), April 18, 2011, Geospatial Information Authority.

[4] Atomic Energy Basic Act, 1956.

[5] Act on the Regulation of Nuclear Source Material, Nuclear Fuel Material and Reactors, 1957.

[6] Act on Prevention of Radiation Disease Due to Radioisotopes etc., 1958.

[7] Designated Radioactive Waste Final Disposal Act, 2000.

[8] Electricity Business Act, 1964.

[9] Order for Enforcement of the Act on the Regulation of Nuclear Source Material, Nuclear Fuel Material and Reactors, 1957.

[10] Order for Enforcement of the Act on Prevention of Radiation Disease Due to Radioisotopes, etc., 1957.
[11] Rules on Building and Operating Practical Electricity Generating Nuclear Reactors, 1978.
[12] Rules on Business of Storing Spent Fuel, June, 2000.
[13] Act for Establishment of the Japan Atomic Energy Commission and the Nuclear Safety Commission, 1955.
[14] Basic Act on Disaster Control Measures, 1961.
[15] Nuclear Emergency Response Manual, 2012.
[16] National Police Agency (in Japanese): http://www.npa.go.jp/archive/keibi/biki/higaijokyo.pdf (accessed 20.08.13).
[17] Reconstruction Agency (in Japanese): http://www.reconstruction.go.jp/topics/main-cat2/sub-cat2-1/20140527_kanrenshi.pdf (accessed 20.08.13).
[18] Fukushima Prefecture Disaster Recovery Office, "2011 Tohoku Area Pacific Offshore Earthquake Damage Report (#1245)" as of August 5, 2014 (in Japanese). http://www.pref.fukushima.lg.jp/sec/16025b/shinsai-higaijokyo.html (accessed 20.08.13).
[19] District population as of October 1, 2010 from "Annual Fukushima Statistics 2012." (in Japanese) http://www.pref.fukushima.lg.jp/sec/11045b/35366.html (accessed 20.08.13).
[20] http://www.nsr.go.jp/english/nuclearfacilities/fukushima1/ (accessed 20.08.13).
[21] http://www.nsr.go.jp/english/nuclearfacilities/fukushima2/ (accessed 20.08.13).
[22] http://www.nsr.go.jp/english/nuclearfacilities/kashiwazaki/ (accessed 20.08.13).
[23] TEPCO, general data on nuclear power plants in operation. http://www.tepco.co.jp/en/corpinfo/ir/tool/illustrated/pdf/illustrated201306-e.pdf (accessed 20.08.13).
[24] Investigation Committee on the Accident at the Fukushima Nuclear Power Stations of Tokyo Electric Company, Interim report, December 2011, p. 18. http://www.cas.go.jp/jp/seisaku/icanps/eng/interim-report.html (accessed 20.08.13).

The Fukushima nuclear power plant accident: the main sequence of events

2

M. Fuchigami[1], *N. Kasahara*[2]
[1]Komatsu Ltd, Tokyo, Japan; [2]The University of Tokyo, Tokyo, Japan

2.1 Introduction

This chapter explains, primarily from a technical viewpoint, analyses of the causes and progress of the Fukushima-1 nuclear accident for 1 week starting from March 11. We start the chapter with an overview of primary facilities involved with the accident and follow it with the main part of the chapter, "overall process from earthquake to station blackout (SBO)" and "processes with each unit." Our explanations follow the timeline. The chapter then discusses the hydrogen explosions and water injection to the spent fuel pools. We then close the chapter with a technical evaluation of the possibility of avoiding the accident and supplemental basic material for better understanding of this accident.

2.2 Outline of primary facilities at the Fukushima plant

2.2.1 *Reactor building facilities and components*

Figure 2.1 shows the cross section of the reactor building and the turbine building. The reactor building housed the nuclear reactor and had five floors and a basement. The building reached about 45 m above ground and contained the reactor pressure vessel (RPV), containment vessel (CV), spent fuel pool (SFP), and other equipment. Many of the pumps for emergency cooling systems were located in the basement of this building. The cooling facilities for Unit 1 isolation condensers (ICs), in contrast, were on the fourth floor, higher than the RPV because the IC operated with natural convection. The safety relief valves (SRVs) were within the CV, whereas the vent valves were outside it. The turbine building contained the turbine generator, main condenser, and most of the electrical power supply facilities. The turbine building basement housed six of the eight emergency diesel generators (D/Gs) for Units 1 through 4. They were destroyed with this accident. Most of the switchboards were on the first and basement floors. It is well known that the direct cause for worsening the accident was the submergence of these switchboards under the tsunami waves. We will briefly discuss each primary facility that played a role in this accident. Each facility is described in normal state except when otherwise noted to discuss the behavior during the accident.

The 2011 Fukushima Nuclear Power Plant Accident. http://dx.doi.org/10.1016/B978-0-08-100118-9.00002-4

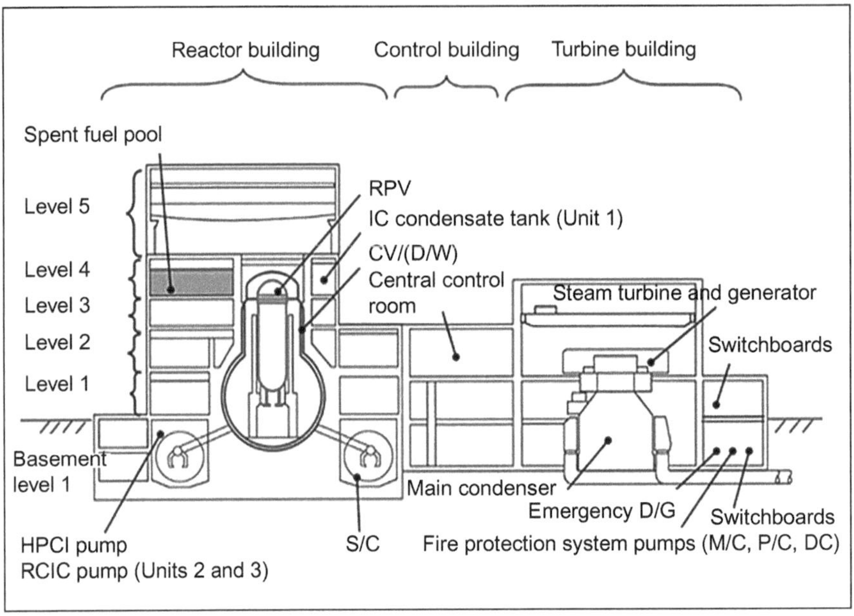

Figure 2.1 Cross-sectional view of reactor building and turbine building.
Note: The black circles indicate the floors where the primary facilities are located and not their exact locations.

Drywell (D/W)

The D/W, shaped like an inverted lightbulb, and the suppression chamber (S/C), compose the CV. The D/W and S/C are connected with eight thick vent pipes. When gas travels to the S/C from the D/W, it bubbles up in the S/C pool. The name *drywell* comes from the absence of water in contrast to the S/C, also called the wet well.

Suppression chamber (S/C)

The S/C is the toroidal vessel that makes the lower part of CV. The Unit 1 S/C contains as much as 1,750 tons of water and those of Units 2 through 4, 2,980 tons. After accidents like a pipe rupture or SRV activation, high-temperature steam discharges into the S/C, and the cooled steam condenses to water to control the CV pressure from rising (Figure 2.2). S/C is also called the pressure S/C. The chamber also supplies water to emergency systems like reactor core isolation cooling system (RCIC) or high-pressure coolant injection system (HPCI) and is also called the wet well or torus.

Safety relief valve (SRV)

Safety relief valves activate to reduce the RPV pressure when it exceeds the allowable value (see Section 2.11.3 for details). Each reactor has eight SRVs (four in the case of Unit 1), which share a number of functions among them.

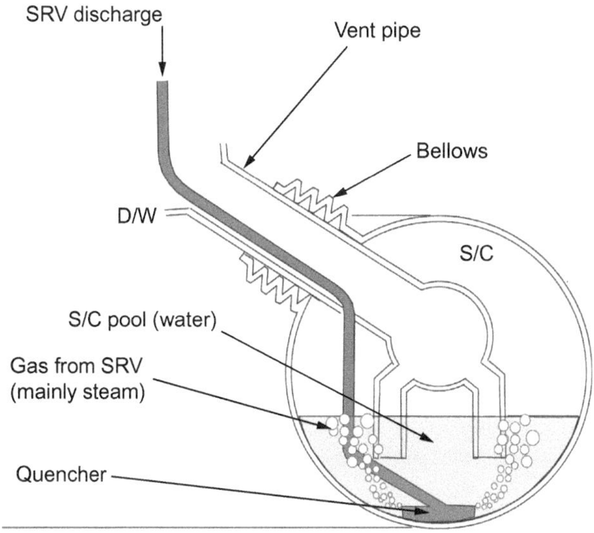

Figure 2.2 SRV discharge to S/C.

Vent valve

Vent valves open to vent the steam inside the CV to the outside in case of excessive CV pressure during a severe accident. The vent line has two different types of valves that both had to open for venting.

Diesel-driven fire pump (D/DFP)

D/DFP are fire pumps driven by diesel engines. Each reactor has one unit to back up the electrically driven fire pump.

2.2.2 Electrical power facilities

A nuclear power plant during normal operation uses the electricity it generates itself; however, when the generation stops, it uses external power sources. If the external power sources are unavailable as well, it will rely on emergency diesel power generators. Figure 2.3 shows the Fukushima-1 NPP electricity system.

Metal-clad (M/C) switch gear

This is a switchboard for high voltage of 6900 V. The power drives large-size equipment like the seawater pumps and condensate pumps. M/C, as Figure 2.3 shows, is the primary equipment of the electrical power systems.

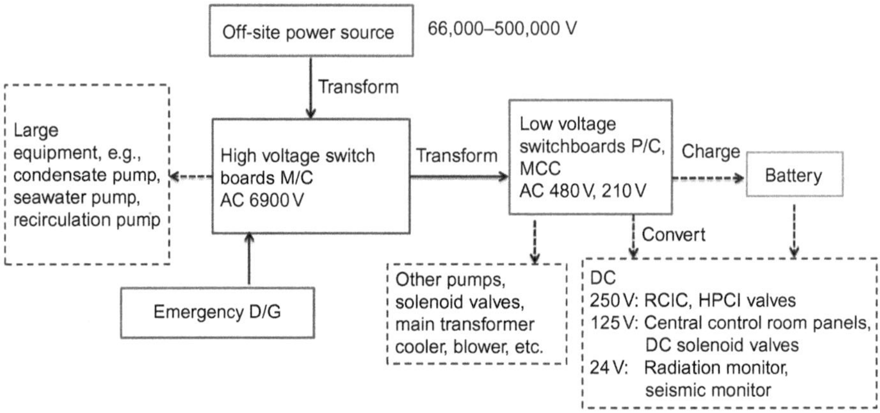

Figure 2.3 Fukushima-1 NPP electricity system. *Note*: M/C and P/C both have two system for regular and emergency use.

Power center (P/C)

P/C is the switchboard for distributing AC low-power voltage of 480 V. Many machines throughout a power plant are driven with power from this switchboard.

Direct current (D/C)

D/C systems transform the AC power from the P/C to D/C and charge batteries as needed. D/C supplies power for control and instrumentation and driving D/C motor-operated (M/O) valves and is the most important power source for power plants. D/C, backed up by batteries, was believed to have low risk in its loss; however, this accident destroyed many of them. There are three types of D/C: 250, 125, and 24 V.

2.2.3 Cooling systems

Systems to cool the core of a nuclear power plant apply to statuses from "normal operation with power generation" to "severe accidents that may lead to core damage." Their functions are most important when the reactor is stopped, either for scheduled maintenance or unexpectedly in the case of an accident. Figure 2.4 shows all the nuclear reactor cooling systems. Fundamental knowledge about these systems is important in understanding the progress of the accident described in the remainder of this chapter. Next we will explain the minimum knowledge necessary for that purpose.

Core cooling systems during normal operation

Steam with high temperature and high pressure from the nuclear reactor turns the turbine, is routed to the main condenser, and returns to the reactor as water. Nuclear fission in the reactor core during normal operation generates a large amount of heat. About one-third of the heat turns into electricity, and the main condenser has to

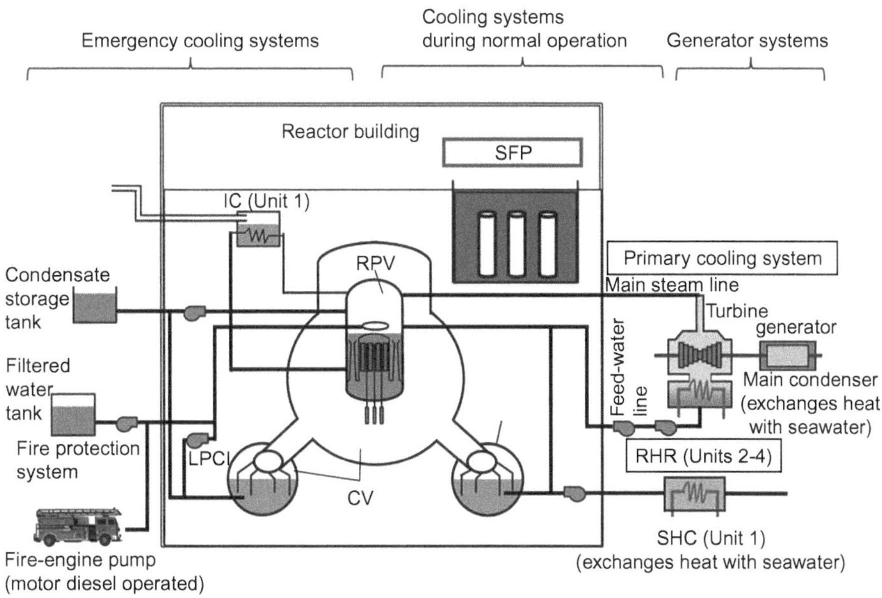

Figure 2.4 Reactor core cooling systems (Unit-1). The unit has three categories of cooling systems: main cooling systems for power generation, cooling systems for normal shutdown, and emergency cooling systems.

remove the remaining two-thirds so it has a large cooling capacity. The heat is then removed at the heat exchanger and discarded into the sea.

Core cooling systems during normal shutdown (including after emergency shutdown [SCRAM])

Nuclear fission stops after normal shutdown for periodic maintenance or emergency shutdown following causes like an earthquake. The core, however, continues to generate decay heat. Without cooling, the reactor has no place to discard its heat, so the residual heat removal system (RHR), although with a capacity smaller than the main condenser, starts running to cool the core.

Emergency cooling systems

Normal shutdown is a condition of normal operation. SCRAM that might be caused by an earthquake is also a normal operation designed to take place in such emergency situations. For preparation against abnormal emergency events like RHR inoperability, which was experienced with this earthquake, or loss of coolant accident (LOCA) (see Remarks) from a piping failure, the reactor is equipped with a number of emergency cooling systems. Some may require electricity for their operation, but others like isolation condenser (IC) with Unit 1 and RCIC with Units 2-4 work without electrical power. High-pressure systems can inject water even when the reactor is at high normal operation pressure of about 7 MPa, whereas the lower pressure systems can only do so when the reactor pressure is at 1 MPa or less. These emergency cooling systems offer a variety to counter various situations.

Remarks

LOCA: An accident caused by, for example, large breaks in the piping with rapid loss of cooling water. Those in the nuclear power community immediately think of LOCA on hearing about an "emergency" situation. Engineers in the past had concentrated on preparation against LOCA after an earthquake.

Isolation condenser (IC)

IC is a condenser for the emergency situation of core isolation. This type of core cooling system was only present with Unit 1. It operates even with high reactor core pressure with natural convection without any power sources (Figure 2.5). The system can withstand an extended operation by supplying cooling water to the condensate tank.

Reactor core isolation cooling system (RCIC)

Units 2–4, instead of IC, have high-pressure isolation core cooling systems. Steam from the reactor core drives a turbine to turn a pump; thus it operates even under loss of AC power. It is designed to run for about 8 h. Its start-up and control requires D/C power, and Unit 2, which lost D/C power with this accident, had no control over RCIC.

High-pressure coolant injection system (HPCI)

HPCI is an emergency cooling system present with all units. This system injects water into the reactor core even when it is at high pressure. Similar to RCIC, the steam in the RPV drives the system. It has a large water injection rate giving it the position of the most powerful cooling system to counter severe accidents like LOCA. During this accident, only the system in Unit 3 worked.

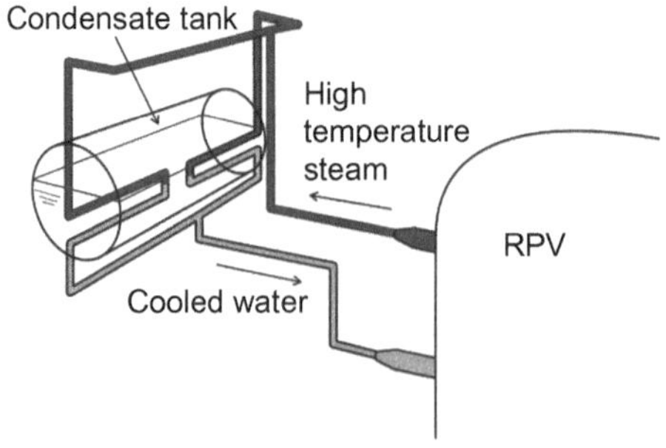

Figure 2.5 Conceptual sketch of the IC. High temperature steam circulates naturally to be cooled in the condensate tank to turn into water and flows back into the reactor core.

Fire protection

Fire systems for nuclear reactors are prepared to provide core cooling if necessary. The system can only inject water when the reactor is at about 0.5 MPa; thus the reactor pressure has to be lowered before the system can operate. During this accident, however, almost all fire pumps within the turbine buildings were inoperable, and fire engines injected water from outside the turbine buildings through water injection inlets on the outer walls to cool the reactor core.

2.3 The sequence of events from earthquake and tsunami to station blackout (SBO)

2.3.1 *Events immediately after the earthquake*

March 11, 2011, about 14:46: magnitude 6-strong earthquake hits

When the earthquake hit, Units 1 through 3 were in normal operation, and 4 through 6 were in scheduled shutdown. Upon the earthquake attack, Units 1 through 3 automatically SCRAMed, and the operators switched the electric power to the site to an external power source. Almost at the same time of the switching, external power was lost (Figures 2.6 and 2.7), and the emergency D/Gs (Figure 2.8) kicked in automatically. As part of the fail-sale function, the main steam isolation valves (MSIVs, valves to close the main steam line to the steam turbine) automatically closed. The series of processes following the SCRAM proceeded normally.

Figure 2.6 "Yorunomori Line" Steel Tower that collapsed. Often quoted to show loss of external power; however, these lines were for Units 5 and 6, not for 1–4 that had severe accidents. Also, loss of external power was caused by failure of breakers, and whether the collapse of this Yorunomori line tower contributed to loss of power remains unknown.

Figure 2.7 Breaker that fell in the switchyard (for Units 1 and 2, Okuma line 1L).

Figure 2.8 Emergency D/G. Output is about 8100 kVA (Units 2-4), and size is about that of a 50,000 ton class tanker.

March 11, about 14:50: Unit 2 RCIC started manually

With the SCRAM, the feedwater pump for Unit 2 stopped, and following the manual, the operator manually started the RCIC with remote operation (turning the switch in the central control room). The RCIC, however, after about a minute of running, automatically stopped because it detected a high water level in the nuclear reactor. The high water level was due to the injection capacity of 95 ton/h with the RCIC, which had started injecting water from the external condensate storage tank to the RPV. Eleven minutes later, the operator restarted the RCIC remotely with his eyes on the reactor water level.

March 11, 2011, 14:52: Unit 1 IC started automatically

The MSIV closure with Unit 1 resulted in a reactor pressure rise, which triggered the IC to automatically start. At about 15:03, however, the operator, worried about the rapid reactor temperature drop (see Remarks), remotely stopped the IC. From that time until 15:30, the operator started and stopped the IC three times, controlling the reactor pressure at 6-7 MPa.

Remarks

The manual cautions against rapid temperature drop and states to keep it at 55 °C/h or slower to prevent damage to the equipment from thermal stress.

Unit 1 had two IC systems, A and B. The capacity of one IC was sufficient for cooling during core isolation; thus, at the time of the accident, the operator worked with the A system only. To operate an IC system intermittently, the operator would open and close one of its four valves and keep the other three open (see Figure 2.24).

March 11, 15:05: Unit 3 RCIC started manually

In the same manner as with Unit 2 the operator started the Unit 3 RCIC remotely. At about 15:25, however, on detecting high reactor water level, the RCIC stopped automatically. At this time, the reactor pressure values with Units 2 and 3 were too high, and the pressure relief functions of the SRVs had already started to discharge RPV steam into the S/C.

March 11, about 15:27: first tsunami wave

The first tsunami wave arrived; however, it was low enough to not cause damage that led to the SBO.

March 11, about 15:35: second tsunami wave

The second tsunami wave reached the coast. The inundation height reached 11.5-15 m and even 17 m at some places. First, the seawater pumps for emergency D/Gs, located near the coastline at an elevation of 4 m above sea level, suffered water damage (Figure 2.9). The details of the damage are unknown, but they lost all functions.

The plant main area with the reactor building and turbine building was at 10 m above sea level; thus, they suffered a maximum inundation of 7 m. The water entered the buildings through doors and air ventilation holes, and most facilities in the basement of the turbine building, including the switchboards, were submerged. This wave caused the loss of almost all electric power sources and was the starting point of this severe accident. The plant response headquarters were set inside the seismic isolation building in a room with no windows and no cameras to look outside; thus, for a time after the SBO, headquarters staff could not tell that the problems were caused by the tsunami.

One report after another reached headquarters with the news of loss of AC power with all units and even DC power with Units 1, 2, and 4. Headquarters staff was

Figure 2.9 Elevations of Fukushima-1 Nuclear Power Plant Central Area.
Photo: Air Photo Service.

stunned by the scope of a catastrophe beyond any imagined scenarios. The plant manager realized that the situation was beyond any hypothesized severe accidents, but could not decide what immediate action to take. He thus followed regulations and reported the SBO to the ministers and agents.

The states of water submergence were about the same for Units 1 through 3; seawater entered the turbine buildings from the equipment hatch and ventilation holes to flood the basement and the concourse floor between the basement and first floor of the turbine, control, and reactor buildings. The basement of the turbine buildings contained the emergency generators and the regular and emergency switchboards; the control building basements contained the DC power systems (Units 1, 2, and 4); and the reactor building basements were where the pumps and other equipment for the emergency cooling systems of RCIC and HPCI were staged (Figure 2.10).

When the seawater pumps fail, the nuclear reactor loses its heat sink (where to dump excessive heat) to the ocean. The Unit 1 IC was designed to release steam to the atmosphere. Units 2 and 3 with RCIC could transfer the heat to the S/C; however, they had lost their ultimate heat sinks. The Unit 1 IC also failed to function; thus, all units that were up and running at the time of the accident had lost their ultimate heat sinks other than venting the CV to release the heat into the atmosphere.

March 11, about 15:39: Unit 2 RCIC started manually just before the tsunami damage

At about this time, an operator opened the Unit 2 RCIC valves remotely. This action was immediately followed by the SBO. If the opening had been delayed by seconds, the Unit 2 RCIC would have been lost (see Remarks), and situations with Unit 2 would

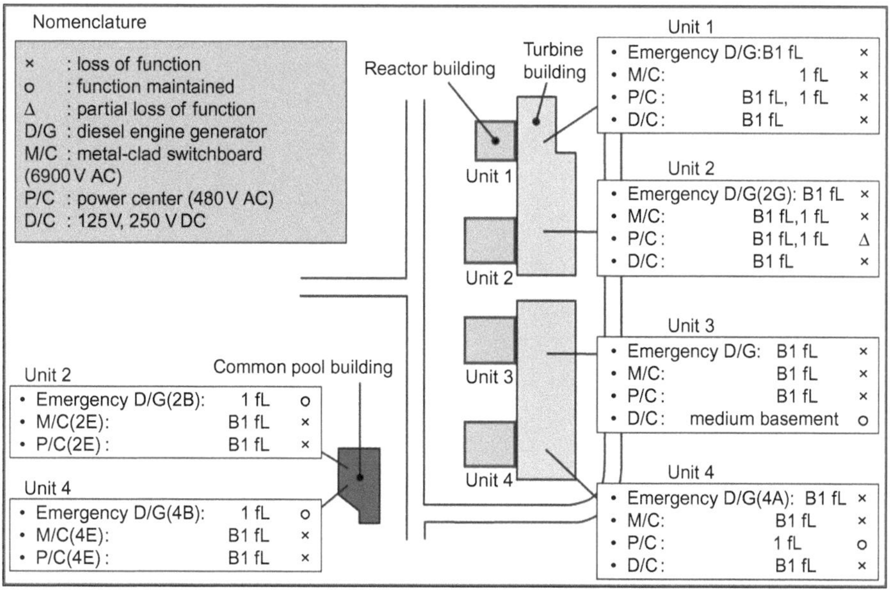

Figure 2.10 Locations of emergency D/Gs and their switchboards.

probably have been worse than the actual trace of events. The runaway Unit 2 RCIC kept running for about 3 days.

Remarks

RCIC control: when the RCIC valves are open, steam from the RPV turns its turbine, and the system runs even without AC power. Opening and closing the valves require 125 V DC, and Unit 2 had lost this DC power (in contrast to Unit 3, which had DC power).

March 11, 15:37-42: loss of all AC power

Except for the Unit 6 air-cooled emergency D/G (6B), all AC power sources for Units 1 through 6 were lost. Units 1 and 2 lost their DC power as well. Unit 3 had its DC power on the concourse level between the basement and first floor. The unit was covered with water, but managed to retain its function; Unit 3 allowed operation of its RCIC and HPCI for a few days.

2.3.2 Loss of electrical power

Each of Units 1 through 4 had two sets of emergency D/Gs making the total count eight. The B-system of Unit 2 and B-system of Unit 4, added in 1999 to enhance plant safety, were located on the first floor of the common spent fuel pool building at 10 m above sea level. All the other six were in the basements of the turbine buildings.

The two D/Gs in the common pool building were covered with water; however, they were not submerged and survived the flooding. These two were air cooled; thus,

Figure 2.11 Switchboards that were submerged under water. M/C in Unit 1 turbine building. The watermarks are at shoulder level.

they were not affected by the failure of destroyed seawater pumps near the coast. The fatal problem was not the loss of D/Gs themselves; it was the sub-mergence and loss of the switchboards (Figure 2.11). The two emergency D/G systems in the common pool building had D/Gs on the first floors, but their switchboards were in the basement. The two, thus, similar to the other six D/G systems, lost their functions, and Units 1 through 4 lost all their AC power.

The scenario of SBO is often described as "the earthquake destroyed the external power source, and then the tsunami submerged the emergency D/G to cause the total loss of AC power"; however, this description is wrong. Among the switchboards of Units 1-4, all M/C switchboards and most of the P/C switchboards were submerged and had lost their functions. For this reason, even if the external power had reached the entrance of the plant switchyard, the state of loss of all AC power during the initial stage of the accident would probably have been the same. Unit 6, in contrast, had one emergency D/G available, and its M/C was functional. Furthermore, it could spare the electricity to Unit 5, and these two, Units 5 and 6, did not have to face SBO. The definite difference with Units 1 through 4 was not the emergency D/G but the high-voltage M/C switchboard.

The electric systems were designed so that the Units 1 and 2 pair and the 3 and 4 pair could spare electricity for each other. Therefore, if the switchboards were available, the two emergency generators that survived the tsunami attack could most probably have supplied the minimum flow of electricity to prevent severe core damage.

2.4 Possible damage caused by the earthquake

2.4.1 *Primary facilities in the reactor building*

The Investigation Committee Report (Interim [1] and Final [2] Reports by the Investigation Committee on the Accident at Fukushima Nuclear Power Stations of the Tokyo Electric Power Company [TEPCO]) questioned the possibility of

Table 2.1 Findings of earthquake-induced damage to primary facilities

Facility	Unit	Findings
RPV	1-3	(1) RPV (2) Instrumentation piping feed-through and their connection to RPV (3) SRV flange gasket and other attachments to RPV have *no possibility* of *damage to affect the containing functions* *explanation: (1), (2), and (3) are three subjects that "had no possibility of damage to affect the containing functions"*
CV	1-3	(1) CV (2) CV flange, electrical wire feed-through, equipment hatches, personnel airlocks, instrumentation piping and their attachment had *no trace* of *damage to largely affect the containing functions* *explanation: (1) and (2) are two subjects that "had no trace of damage to largely affect the containing functions"*
IC	1	Could not find *traces of breakage* in piping and condensate tank to *have impaired the IC function*
RCIC	2, 3	Possibility was *denied* for *severe damage* to *have impaired water injection function*
HPCI	1, 2	Possibility of *damage* to *have impaired the water injection function* was *low*
	3	Possibility was *denied* about *damage* to affect the *water injection function*

earthquake-induced "damage to affect the containing function" during the 50 some minutes from earthquake to SBO. The relevant facilities are shown in Table 2.1, i.e., RPV and its attachments, CV and its attachments, IC and its piping, RCIC, and HPCI (Unit 3 only). The report also confirms that for Units 1 and 2, "HPCIs have low possibility of suffering damage to lose their water injection functions." Table 2.1 summarizes these evaluations.

Remarks

The Investigation Committee Report has carefully chosen expressions about its statements (e.g., in the table "denial" is definite, whereas "low" possibility means the evaluation could not conclude a full denial. In the rest of this book, expressions have distinct meaning with "be," "highly possible," "possible," and "possibility cannot be denied").

The "possibility" the Investigation Committee denied was "damages that disabled the containing functions," in other words, "damages with openings of significant size." This denial does not mean there were no smaller cracks. There remains the possibility that "small cracks caused by the earthquake might have widened later with the SBO." It is another fact, nevertheless, that traces of such sequences have not been identified with the equipment either.

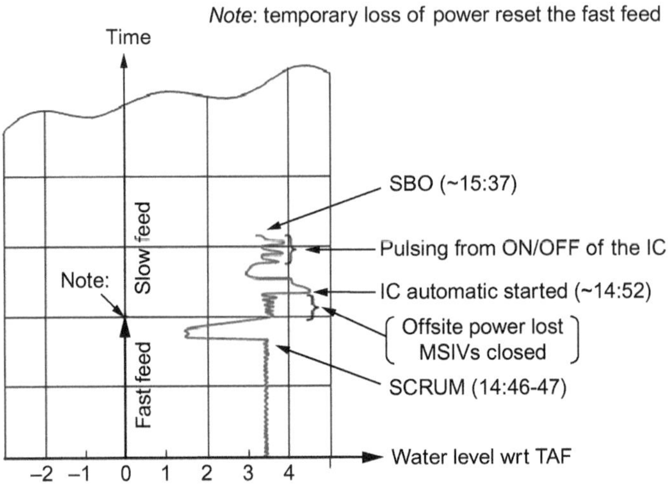

Figure 2.12 Unit 1 reactor water level transition shows that IC repeated ON/OFF as designed.

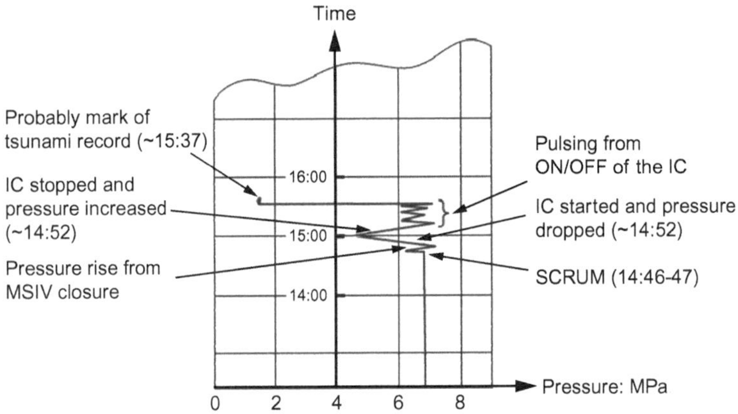

Figure 2.13 Unit 1 reactor pressure transition.

The main reasons for claiming that there was no serious earthquake-induced damage to the primary facilities follow (e.g., for Unit 1):

(i) RPV

Figures 2.12 and 2.13, respectively, show the traces of Unit 1 RPV water level and pressure from before the SCRAM to SBO. Transitions in these charts support the sequence of SCRAM → MSIV closure → IC start → IC stop. If the RPV or piping that penetrated it (e.g., IC piping) had large cracks, the charts would have shown a rapid drop in water level or pressure. We could not find such traces. Also, signals of LOCA (trips on large loss of cooling water) to trigger HPCI, emergency D/G, or core spray systems could not be found either.

Furthermore, later at 20:07 on March 11th, the RPV recorded pressure of 6.9 MPa. This value in the normal range over 5 h after the earthquake backs the low possibility of earthquake-induced RPV damage that impaired its containing function. The pressure records were directly read by a worker from a mechanical gauge on the second floor of the reactor building and thus were judged to be highly reliable with hardly any reasons for misreading.

Moreover, the Containment Atmospheric Monitoring System (CAMS) marked the lowest level of radiation up to the point of SBO for both the D/W and S/C, and there were no signs of leakage to the CV.

All the preceding records show that the RPV maintained its containing function up to the point of SBO.

(ii) CV

Records of D/W pressure and temperature, S/C pressure, and water temperature all show no sign of CV damage to largely impair the CV containment function before the tsunami attack.

In addition, during this time period, the standby gas treatment system (SGTS; see Remarks) operated to exhaust the reactor building air from the exhaust tower; however, the exhaust tower radiation monitor did not record a significant rise in radiation. This fact also supports that there was no significant radiation leakage from the CV to the reactor building, thus, the CV's containing function was intact.

Remarks

SGTS: SGTS exhausts gas with radioactivity in the reactor building to the outside after filtering.

So far we have explained that Unit 1, and Units 2 and 3 (although with the difference of IC and RCIC), for similar reasons, did not receive possible large damage caused by the earthquake to the primary facilities of RPV, CV, RCIC, HPCI, and their attachments to impair the containing function.

2.4.2 Other facilities

Starting with Unit 1, the accident response repeatedly had to face the lack of compressed air. If the AC power was available, the stationary compressors could have provided that function, but they were unavailable with this accident.

All the vent operations required compressed air; however, the air bottles stored in the buildings were not sufficient, and the workers had to rely on portable compressors or borrow them from contractors of Fukushima-2. All these compressors gathered for the urgent situation, however, did not have sufficient capacities, which caused delays in the vent operations. In case of Unit 2, the venting, after all, most likely had failed. Whether the failure was due only to the capacity of compressors or if there were leakages from the compressed air piping is still unclear. In any case, we have identified a problem that seismic class S venting facilities required seismic class C air piping (Figure 2.14).

The fire protection systems piping outside the buildings were also only classified as Class C. During this accident, workers connected fire engine hoses directly to the turbine building water inlet; however, the original accident management (AM) response (see Remarks) had specified to use the filtered water tank as the source. The distance from the filtered water tank to the primary facilities was over 500 m, and after the earthquake, some fireplugs were seen shooting out water. Plant manager Yoshida stated that he thought "diesel-driven fire pumps (D/DFP) relying on filtered water tank as a water source was unreliable."[8].

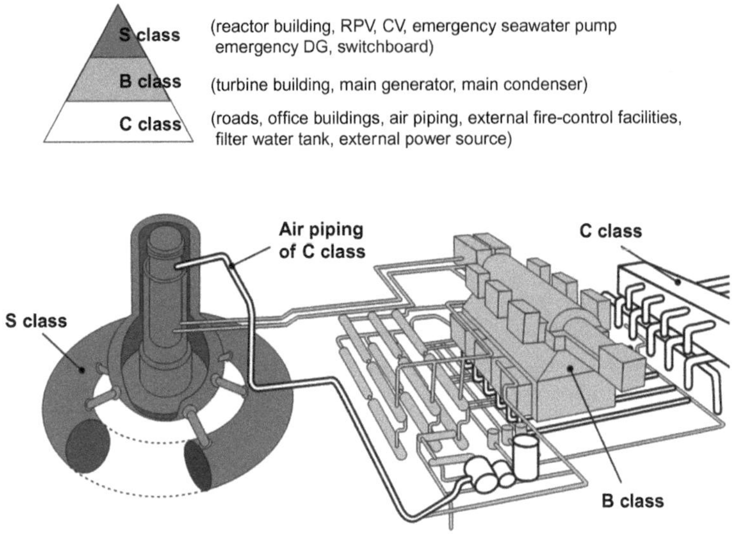

Figure 2.14 Aseismic classification of plant facilities.

Remarks

AM: responses, prepared in case of events that exceeded the design expectation range, to lower the severity of the accident. Each electricity company was instructed to independently prepare them by about 1995.

The auxiliary facilities, if they are needed at times of severe accidents, need reevaluation of their seismic classification. The reevaluation should look at severe accidents caused by external events like earthquakes and also tsunami attacks, mudslides, airplane crashes, and terrorism attacks with discussion on the overall response of the plant.

2.5 The condition of Unit 1 after SBO

2.5.1 March 11, up to 23:50 when abnormal CV pressure was recognized

March 11, about 15:37: IC isolation valves closed with fail-safe function

Unit 1 lost all its electrical power including DC. The central control room was pitch dark, and the loss of DC power stopped displays of all instruments. The most critical reactor water level and pressure became unknown.

The isolation condenser (IC) was in normal operation with the operator turning it on and off. The loss of DC power, however, caused the fail-safe logic to send "close" signals to all four valves. High-temperature steam from the reactor stopped feeding to the IC to be condensed into water to run back to the reactor. The cooling function of the IC was almost completely lost. The Unit 1 IC was designed that on failure (an abnormal condition), all flows were stopped to prevent possible radioactivity leakage and to contain the pressure vessel. This concept, however, had the contradiction that shutting off the IC cooling function could lead to a more dangerous situation.

Around this time, Plant Manager Yoshida was not sure whether the Unit 1 IC and Units 2 and 3 RCIC were properly operating, but at the same time was not aware of the Unit 1 IC stoppage with the fail-safe function.

Unit 1 lost both the DC for control and AC for driving power, and when the "valve close" signals were sent, the driving power to make those actions (480 Vac for the two valves inside the CV and 125 Vdc for the two outside) were also lost. Thus, whether the valves actually closed or stopped midway is unclear. The fact of rise in radioactivity recorded only 2 h after the SBO at about 18:00, and other evidences confirm that the IC was not functioning after the SBO (see Section 2.5.3 for details).

Remarks

As noted earlier, the fail-safe design to close the valves came from the design philosophy to "prevent radiation leakage by all means" by "closing all valves." This philosophy contradicts the requirement to "keep the valves open" on a severe accident to the first priority of keeping the core cooled. The fundamental choice to make here was whether to "counter even the smallest accident" or to "accept small damage to prevent catastrophic ones." The ruptured disk we will discuss later had a problem of the same nature. The RCIC systems for Units 2 and 3 were designed so their isolation valves remain "as is" on loss of DC power.

Another root cause for failing to recognize the IC stoppage, in addition to the preceding design problem, was the lack of education and training for the hypothetical situation of loss of DC power. Furthermore, for this March 11 accident and the hours that followed, none of the people that were responding to the accident, in the plant response headquarters, main office response headquarters, Nuclear and Industrial Safety Agency (NISA), or Nuclear Safety Commission (NSC), recognized the IC stoppage. One of the reasons why the operator on duty did not recognize the fact was because no one had ever experienced the IC in operation. The best knowledge that anyone had was having heard from former operators that when IC was in operation, steam from the condenser cooling water would blow out horizontally from the exhaust openings (the so-called pig snout) to produce lightning-like static and large thunder-like sounds (Figures 2.15 and 2.16). The reason why no one had experience of the event was because Unit 1 was the oldest among those with MARK-I containment. There was only one other reactor, Tsuruga-1 of The Japan Atomic Power Company (JAPC), that had adopted IC for its cooling system in the event of reactor core isolation.

The Unit 1 operator was puzzled by the small amount of steam from the IC condenser, but did not have a good understanding of the situation. He thus switched Valve 3A from "Open" to "Close" and "Open" again three times between 18:00 and past 21:00. Because the three valves other than 3A remained (almost) shut due to the fail-safe function, operating Valve 3A was meaningless. As a result, Unit 1 was in an abnormal situation with no water injected into the core until water injection from a fire engine started at 4:00 the next morning. This is how the core melting progressed rapidly with Unit 1, for which water injection had stopped at only 1 h after its SCRAM when the decay heat was still high.

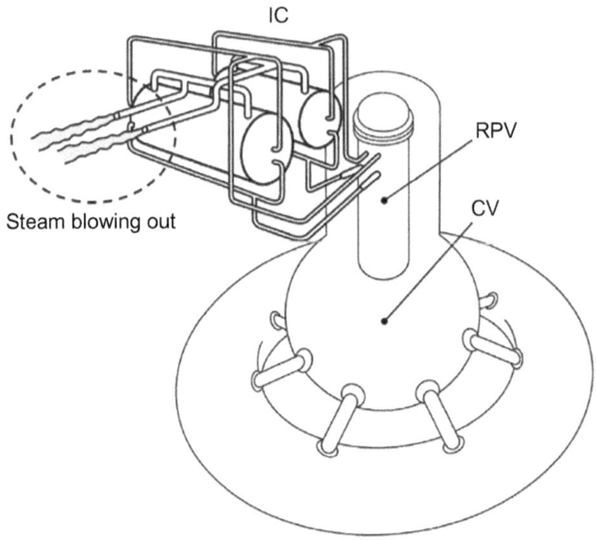

Figure 2.15 IC in operation. Cooling water in the condenser tank blows out as steam from the exhaust holes, so-called "pig snout."

Figure 2.16 "Pig snout." Only Unit 1 has two pipes sticking out from the upper part of the building wall. These pipes are not visible from the central control room.
© Takuo Kawamoto.

March 11, 16:42: drop in water level detected

At about this time, operators recognized that the indicator of the wide range water level monitor (one of the three types of reactor water level monitors) was working, but the reason for this sudden recovery remains unknown. The indicator at first was pointing at "−900 mm," and 14 min later after showing "−1500 mm," it went off scale. During this time period, the water level monitor was probably working normally, and the plant response headquarters, given information about the rapid drop in water level, estimated the time for the water level to drop to top of active fuel (TAF) to be an hour; however, in the midst of all confusion, headquarters did not recognize the IC stoppage.

March 11, 17:12: use of water injection from fire engines evaluated

Under SBO, the AM procedure designates the only alternative method for water injection other than IC to be D/DFP driven water injection. The water source in that case was the "filtered water tank (see Remarks 1)." Plant Manager Yoshida was worried about piping from the filtered water tank to the reactor building. He then ordered evaluation of water injection from a fire engine, although it was not prepared as an AM procedure. On March 12th and later, this water injection from fire engines turned out to be the lifeline. At this time there were three fire engines on-site; however, one was unusable with damage from the tsunami, and another was parked near Unit 6, with its way to Unit 1 blocked by tsunami rubble. That left only one engine available for the purpose (see Remarks 2). To make things worse, alternative water injection with a fire engine was not written in the AM procedure, and no team was assigned to take responsibility for the action. The plant manager had given orders but no progress was made until before dawn on the next day, March 12th.

Remarks 1

Filtered water: simply filtered water without radioactivity. The water was stored in two large 8000-ton tanks to share within the plant.

Remarks 2

The number of available fire engines gradually increased; the one near Units 5 and 6 arrived before dawn on March 12th, and three arrived from off-site during the morning on the same day.

March 11, 17:30: doubts about the operation of IC

Some operators on duty in the central control room had doubts about the operation of IC. In preparation for alternative water injection, the operator checked the D/DFP start-up and put it on standby. Later, at 20:50, he started the D/DFP engine to prepare for the change of situation.

Later in the afternoon, four 6 V batteries from a neighboring company and two 12 V batteries from large buses arrived. They were hooked up to the back of the control panel in the central control room to start monitoring water levels of Units 1 and 2.

March 11, 17:50: first abnormal radioactivity detected

A worker approached the reactor building to check the water level of the IC condensate tank; however, near the double door (see Remarks), radiation dosage read as high as 2.5 μSv/h, and he had to retreat. This high-dose reading was much smaller compared to later large readings. However, it was out of normal range, and probably at this time, part of the fuel was exposed above the top of cooling water.

Remarks

Double door: This double door was at the entrance to the reactor building from the turbine building.

March 11, 18:18: IC valves 2A and 3A closed

In the central control room, the operators realized the IC lamps that had been off were illuminated showing "Valve 2A and 3A Closed." At this time the operator on duty thought that all IC valves might be closed with the fail-safe function, but hoping they were not, he remotely opened valves 2A and 3A. He then reported to the response headquarters about the opening. The operator left the central control room to visually check the steam ejection from the reactor building. The operator only saw a small amount of steam at first, and then no more steam followed the initial exhaust.

Remarks

Valves 1 and 4 inside the CV could not be opened with the manual handles located inside the CV because the CV could not be entered. AC power for remote operation was also lost at this time; thus, it was impossible to open these two valves.

March 11, 18:25: IC valve 3A remotely closed

Seeing the small amount of steam from the IC, the operator who had doubts about the IC function began to worry that "steam not blowing out from the IC may be a sign of low condenser cooling water level, and keep running the IC may cause damage to the IC piping," [9,10] and at 18:25 he remotely closed Valve 3A. This operation, however, was not reported to the response headquarters.

March 11, 20:07: RPV pressure acknowledged

A worker entered the reactor building and found the RPV pressure at 6.9 MPa. Around this time, the SRV might had been open with its safety function because the Unit 1 SRVs were set to open at 7.7 MPa and remain open until the RPV pressure drops to about 6.9 MPa. Up to this time, however, workers were allowed in the torus in the basement of the reactor building; thus, a large amount of radioactivity release through the SRVs to the S/C had been unlikely. Within the reactor, therefore, a certain amount of fuel damage at the time cannot be denied; however, it is natural to assume that the core had not reached significant core damage to release a large amount of radioactivity.

March 11, 21:19: water level gauges started to give misreadings

The water level gauge again read "TAF + 450 mm." The water level at this time, however, was unlikely to be above the TAF, so it is most likely that from this point on the water level reading was wrong.

March 11, 21:30: IC reopened

At around this time the operator on duty recognized that the IC Valve 3A indicator lamp showing the green "closed" status was about to go off. The operator was concerned about the risk of being unable to restart the IC due to battery drain. In addition, by this time he had learned that the condenser cooling water would not run out with several hours of operation and reasoned that the small amount of steam from the "pig snout" was not due to the drop of water level in the condensate tank. Although the operator thought that probably the fail-safe function had closed the valves, he also thought there still was a chance of activating the IC and so turned the Valve 3A switch to open. Upon this action, sound of steam release was heard for a while, but then again it faded away. The operator believed that IC was not operating and reported that he had opened Valve 3A to the response headquarters. The response headquarters, not aware of the "close" operation at 18:25, could have found the report puzzling, but no one did.

March 11, 21:51: apparent rise in the radiation dosage

A rise in the dosage led to an order to ban entry into the reactor building.

March 11, 22:00: wrong water level information

Water level measurement gave a value of TAF + 550 mm. This report was a sign of worsening misreading of rising water level despite no reason for the water level to rise.

March 11, about 22:30: large increase in the radiation dosage

An operator on duty tried to enter the reactor building, but in front of the double door, his pocket dosimeter indicated 0.8 mSv within about 10 s, and sensing the risk, he returned to the central control room. This reading was equivalent to as high as about 300 mSv an hour. For information, the standard safe exposure for a person throughout his life is about one-third of this value at 100 mSv.

March 11, 23:50: IC problem finally recognized

The staff measured the D/W pressure using a portable generator from a contractor in the central control room. The reading was extremely high at 0.60 MPa. Plant Manager Yoshida, on hearing this report, finally recognized that the IC was not properly working and that the steam that had leaked into the CV D/W from the RPV had produced the abnormal rise in D/W pressure.

The reason for this D/W pressure rise could have been caused by hydrogen and rare gas, which do not condense to water, discharged into the S/C from the RPV, and with other gas then entered the D/W; however, there was also the possibility of damage on the pressure vessel to let gas escape into the CV.

2.5.2 Further progress

March 12, about 00:06: first order to prepare for venting

Plant Manager Yoshida, upon receiving the report of D/W pressure at an abnormally high value of 0.6 MPa, realized that the IC was not working. With the possibility of even higher pressure, without hesitation, he ordered the power generation team and recovery team of the response headquarters to prepare Unit 1 for venting. At the same time, he ordered to prepare Unit 2 for venting as well in case of worst-case scenario.

March 12, around 01:30: all parties agreed to the venting

At the main office response headquarters, President Shimizu had agreed to the venting of Units 1 and 2. TEPCO also requested approval for the operation from the Prime Minister's Office and NISA and soon received approval from Prime Minister Kan and others.

March 12, 01:48: D/DFP stoppage and evaluation of the alternative fire engine

D/DFP of Unit 1 was running from the day before in preparation for its use; however, it was found to have stopped by this time. The reason for the stoppage is still unknown. The response headquarters reasoned that the only alternative left was to use a fire engine to inject water and started to evaluate that procedure. Water injection to the reactor with a fire engine was a procedure never evaluated before and, thus, faced a number of difficulties. First, none of the TEPCO employees knew how to operate a fire engine. TEPCO had to request a contractor on-site to perform the work, which was outside their contract. The response headquarters did not know where the inlet for injecting water into the reactor was, and extra time was needed to locate a contractor employee with such knowledge. Furthermore, the only nearby water supply was the fire reservoir, which had insufficient water for extended water injection.

Remarks

One of the AM plans was to use D/DFP for reactor cooling water injection in case of a severe accident. The discharge pressure, however, was at 0.5 MPa, which would require depressurization of the reactor vessel. There were no AM plans to use the same fire line for injecting water from an outside fire engine.

March 12, about 02:30: sudden rise of D/W pressure

The D/W pressure was reported to be at 0.84 MPa (abs). Most persons involved would have thought that such a high pressure could cause explosive damage to the CV.

Remarks

The allowable design pressure of CV was 0.427 MPa.

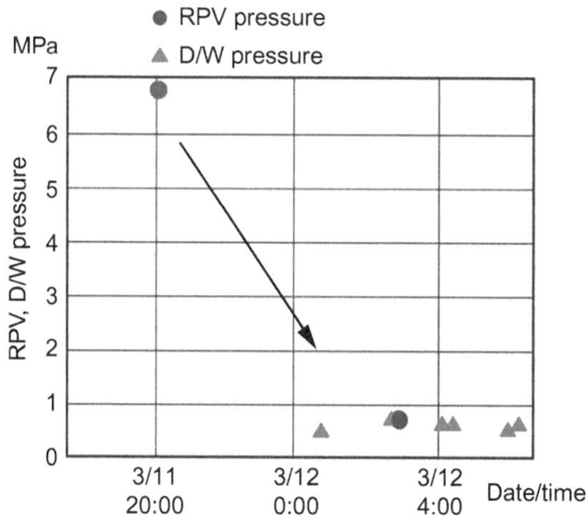

Figure 2.17 RPV pressure and D/W pressure. At 2:45 on March 12th, the first measured RPV pressure had already dropped to the same level with the D/W pressure (possible large-scale damage).

March 12, 02:45: sudden RPV pressure drop

At about this time, the RPV pressure was measured at 0.8 MPa. That was a large drop from 6.9 MPa, measured about 7 h earlier at 20:07 on the previous day (Figure 2.17). No depressurization was made with SRVs during this time, and without doubt, there was a large leakage from the pressure vessel (see Remarks). The rise in the D/W pressure to about the same 0.84 MPa backs this conjecture.

Remarks

SRVs have a safety valve function to automatically release the RPV pressure without operator intervention. This safety function closes the valve at about 6.9 MPa and, thus, could not further lower the pressure. A larger depressurization requires the operator to open the valve to relieve the pressure. No such operation was made during this time, and damage to the pressure vessel was the most likely cause.

March 12, 03:06: press conference about the venting

The Minister of Economy, Trade and Industry, Banri Kaieda, and a TEPCO Executive, Akio Komori, held a press conference to announce the plan to vent. Up to this point, the response headquarters had erroneous optimistic information about the situation with Unit 1 due to misunderstanding the IC status and misreading of the water level monitor. On the other hand, until about after 02:00 on March 12th, because of not knowing the RCIC operation and having no water level reading, officials had thought that Unit 2 was the reactor in most danger. Immediately before this press conference, the government received the news that Unit 2 had its RCIC running, and its water level and pressure were stable. Executive Komori took the stand in the press conference without hearing the news, and when asked, "Which is the first unit to vent?" he gave a confusing reply.

March 12, after 04:00: freshwater injection started

Started to inject freshwater using a fire engine. The water was first supplied from the fire engine tank and then collected from the fire reservoir and poured into the fire engine tank. This intermittent injection amounted to only 1-2 ton every several tens of minutes. At this time, as shown in Figure 2.18, the D/W pressure was going down. Because D/W cooling (D/W spray), S/C cooling (S/C spray), nor venting were not made, the only reason for this pressure drop was leakage from the CV.

March 12, about 04:23: rise in radiation dosage

The radiation dosage at the monitoring post near the main gate (Main Gate MP) rapidly rose to 0.59 μSv/h and to about 1.8 μSv/h at around 05:20 (the reading was around normal at about 0.069 μSv/h at 04:00). Also the central control room for Units 1 and 2 was located on the second floor of the control building between the two units. The dose reading there was high at around 05:00. In addition, the dose measurement was higher where closer to Unit 1. The operators moved toward the Unit 2 side and waited. It was a sign of serious core damage in the Unit 1 reactor located at an elevation higher than the central control room.

March 12, 06:50: concern about not completing the vent

The Minister of Economy, Trade and Industry Kaieda ordered a vent in accordance with regulations. Around this time, the authorities in Tokyo expressed concern about the progress of the vent, and some even speculated that the operators were hesitating in the task. The reality was that a number of difficulties occurred in performing the vent.

March 12, 07:11: Prime Minister Kan's visit to the site

Prime Minister Kan and NSC Chair Haruki Madarame arrived at the scene. Vice President of TEPCO Sakae Mutoh and Plant Manager Yoshida alone from the plant greeted them and explained that they intended to perform the vent at around 09:00.

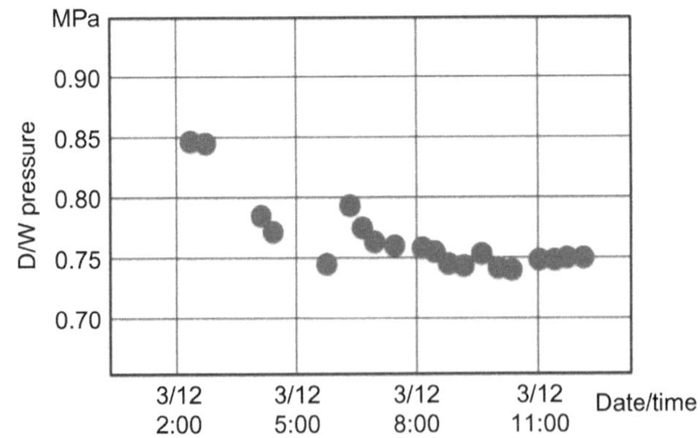

Figure 2.18 Unit 1, D/W pressure in the morning of March 12th. Pressure drop started probably due to leakage from CV.

March 12, 9:15: preparation for opening the vent line

At about this time, a worker went into the second floor of the reactor building and opened the motor-driven vent valve by about 25%. This work, as we described earlier, was necessary to open the vent line.

The radiation level on the second floor of the reactor building at this time was not so high, and a worker could reach there to manually open the valve by turning its handle (Figure 2.19). Then, however, the other air-operated (A/O) vent valve needed compressed air, and insufficient air further delayed the process. In all, it took another 5 h to succeed with the vent.

March 12, about 09:24: opening of the vent line further delayed

At about this time, an operator headed toward the basement of the reactor building to manually open the A/O valve (small). The basement floor at the time, however, had a high radiation level, and the worker had to retreat to the central control room.

Remarks

Only Unit 1 had a manual handle on the A/O valve (small) for the bypass function in place for the large valve. All A/O valves (large) did not have handles with any of the units, and they required air for their operation.

March 12, about 10:17: misconception of vent success

The operator gave up the manual operation and tried to remotely open it from the central control room. He hoped that compressed air in the instrument air system (IA, see Remarks) piping might open the vent valve (A/O small valve). After this trial at about 10:40, the response headquarters received a report of radiation increase at the main gate and monitoring post and thought that the vent had succeeded; however, at around

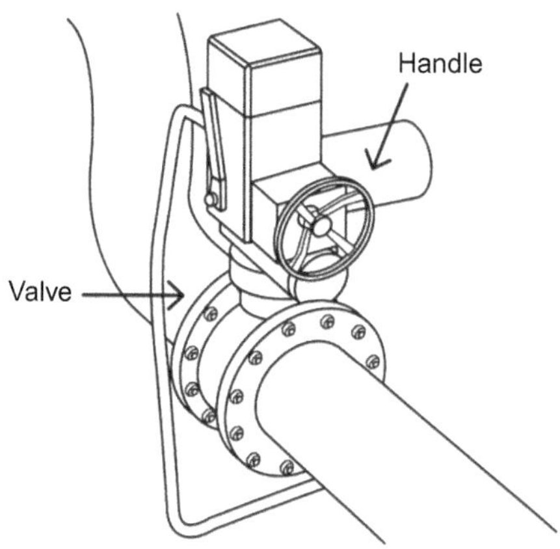

Figure 2.19 M/O vent valve. The valve has a handle for manual turning.

11:15, the radiation level went down again, and headquarters had to correct this earlier conclusion.

Remarks

IA system: instrument air system.

March 12, about 12:00: decision made to switch to seawater

At about this time, Plant Manager Yoshida decided to switch to seawater in case the freshwater was used up, and he ordered the building of a seawater injection line with hoses. TEPCO employees and contractors searched for accessible seawater near the buildings. They soon spotted a large amount of seawater in the reverse flow cleaning valve pit (see Remarks; Figure 2.20).

Remarks

Reverse flow cleaning valve pit: A hollow location where the valves for the cleaning process of reversing seawater into the condenser to wash away deposits and the like were located.

At about this time, TEPCO staff, including VP Mutoh and NSC chief Madarame understood that the seawater injection process was the only choice and had no opposition to it.

Figure 2.20 Reverse flow cleaning valve pit and nearby fire engines (Unit 3 on March 24th).

March 12, about 12:30: vent preparation completed

Around this time, the workers finally obtained a portable air compressor from a contractor. It was needed to open A/O valves. This compressor, however, had a small capacity and had to be set close to the A/O valve and in an area with relatively low radiation. After evaluation, they decided to set it near the reactor building large equipment hatch. The operation, however, took efforts for not having prepared the adaptor and other reasons, but around 14:00, the connection was complete, and the compressor was ready to supply air. The process of opening the A/O valve required opening the solenoid valve on the air piping as well. The central control room, therefore, after starting up the portable compressor, applied AC current from the small generator for the temporary lamps to energize the solenoid.

March 12, 14:50: venting judged to have succeeded

With all the aforementioned work, the D/W pressure dropped to 0.58 from 0.75 MPa, and the image on the TV showed white smoke coming from the exhaust tower. Plant Manager Yoshida judged that the "vent was successful back at about 14:30" and forwarded the information to Ministries and Agents at 15:18.

March 12, 14:53: preparing for seawater injection

At about this time, the freshwater in the fire reservoir was used up, and there was no alternative source of freshwater at hand then. Plant Manager Yoshida thus ordered to inject seawater, and the line construction was almost complete at about 15:30.

March 12, 15:36: hydrogen explosion

Unit 1 had a hydrogen explosion, and five workers in the field were injured (Figure 2.21). For a while, the effect of the explosion had to be evaluated, and until safety was confirmed, the recovery work had to be put on hold. The new seawater injection line was damaged and could not be used.

The three fire engines luckily were found operable. The explosion, however, forced the team to restart the electricity recovery work, which had almost been complete, from scratch. Plant Manager Yoshida was greatly disappointed at the setback caused the hydrogen explosion. The workers at the time had to take cover in the Seismic Isolated Building.

March 12, about 17:20: discussion about the effects of seawater injection

After checking the damage, the field restarted work for seawater injection. At about the same time in the Prime Minister's Office, Prime Minister Kan, Assistant Goshi Hosono, NSC Chief Haruki Madarame, Assistant Manager Eiji Hiraoka, and TEPCO Fellow Ichiro Takekuro had gathered in the prime minister's office. When Prime Minister Kan asked about the effect of seawater injection to the reactor, both Chief Madarame and Fellow Takekuro answered, "Even if it is seawater, we have to put priority on

Figure 2.21 Hydrogen explosion of Unit 1 (15 s after the explosion). The explosion was strong in the horizontal direcion.
Photos © Fukushima Central Television.

injecting water as soon as possible." Prime Minister Kan then asked Chief Madarame, "Will injecting seawater cause re-criticality?" and Chief Madarame answered, "You do not have to worry so much about a possible re-criticality." However, Kan was not satisfied with the answer.

March 12, 19:04: seawater injection started

The seawater injection, whose preparation had been once blocked by the hydrogen explosion, finally started at 19:04.

Fellow Takekuro, who was not aware of the start, strongly requested Plant Manager Yoshida to "Halt the seawater injection because they are discussing it in the prime minister's office now." Yoshida was worried about worsening the situation by stopping the seawater injection and consulted with the TEPCO main office and VP Mutoh at the off-site center. They gave him the same reply: "We have to halt the injection because we do not have the blessing from Prime Minister Kan." Yoshida then judged, however, that he should not stop the seawater injection and whispered to those on duty in a voice low enough to escape detection by speakerphone, "By whatever means, never stop the water injection," and then, in a loud voice that reached all corners of the response headquarters room, he ordered, "Stop the seawater injection."

Later, at about 19:30, discussion resumed in the prime minister's office. Prime Minister Kan, unaware of the process, immediately approved the seawater injection, and Fellow Takekuro lost his chance to explain the situation to Prime Minister Kan. At 20:20, the order to restart seawater injection was given, and the water injection up to that point was recorded as "water injection test" to agree with the time of offical approval.

March 13, past 08:00: radiation increase

For about an hour after this time, radiation levels around monitoring post 4 (MP-4) went up rapidly. The rate, which was 35 μSv/h at about 8:01, rose to 1204 μSv/h at 8:33 (Figure 2.22). This value had 2 more digits compared to the data around the main gate just a few hours earlier.

The reason for this rise was probably not due to the Unit 1 hydrogen explosion or its venting. Most likely it was due to radiation leakage from the CV after core damage of Unit 1, Unit 3, or maybe both.

Unit 1 is estimated to have suffered damage to its CV by about 22:00 on March 11th, and the damage grew by the dawn of March 12th; however, the damage could have expanded more by this time.

March 14, about 01:10: seawater injection halted

At about this time, water withdrawal from the Unit 3 reverse flow cleaning valve pit was disabled, and seawater injection to Unit 1 stopped. The workers then started to build a seawater supply line using a newly arrived fire engine to withdraw seawater directly from the ocean and pour it in the reverse flow cleaning valve pit. The line was completed shortly after 09:00; however, injection to Unit 3 had priority then. To make things worse, Unit 3 had its hydrogen explosion 2 h later, and the fire engine stopped its operation. Water injection to Unit 1, thus, did not restart until about 20:30, meaning that injection to Unit 1 had stopped for over 19 h.

March 14, about 06:00: further radioactivity leakage

Data from CAMS was available by this time. The data showed higher γ-ray values in the D/W than in the S/C from this time on to late afternoon of March 15th. High γ-ray measurement inside the CV despite the already started leakage from the CV means that core damage was still in progress, and levels higher in the D/W than in the

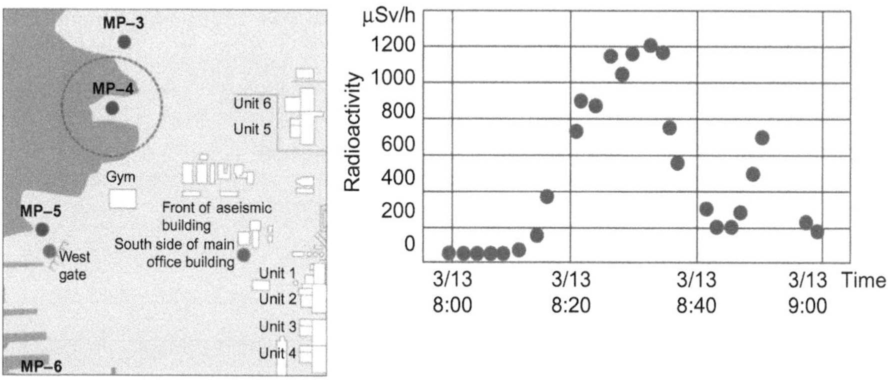

Figure 2.22 Sudden increase of radioactivity at MP-4.
Note: After 08:00 on March 13th, the Unit 3 HPCI had stopped for about 5½ h. Core damage was likely to have started and possible leakage from the CV might have been started by then.

S/C were an indication of a high possibility of leakage from the RPV directly to the D/W (Figure 2.23).

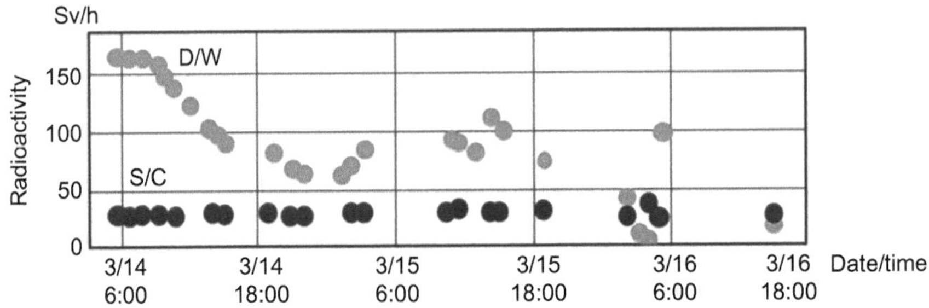

Figure 2.23 CAMS radioactivity data. Strong radioactivity in D/W persisted.

March 14, 11:01: Unit 3 hydrogen explosion

The situation on and after March 15

The CAMS report of high radiation in the D/W during March 15th largely subsided in the night of March 16th, and the situation with Unit 1 had calmed.

There was no water injection into the spent fuel pool of Unit 1 until March 31. The spent fuel there had been cooled for over a year, and its cooling had not been an urgent matter (Section 2.9 will discuss water injection into the spent fuel pools of Units 1 through 4).

Summary of damage to the reactor primary equipment

Table 2.2 shows the damage to the core, RPV, and CV of Unit 1, as they were confirmed by the Investigation Committee. The expressions are based on evidence and have been selected carefully. The locations of leakages from the RPV and CV have not been identified.

Table 2.2 Summary of damage to Unit 1 reactor

	Confirmed damage to Unit 1
RPV	• Natural to believe damage was present from 20:07 on March 11th to 2:45 on March 12th
	• Further damage may have taken place afterward
CV	• Possible damage by 21:51 on March 11th
	• Further damage may have taken place afterward

Damage: damage that impairs the containment function.

2.5.3 Reason to believe IC did not function

Unit 1 had two IC systems, System A and B. Figure 2.24 shows System A only. Each IC system had four valves, and all of them had to open for IC to function. Of the four valves, two (1A and 4A) were inside the CV, and they were driven open and closed with AC 480 V. The other two (2A and 3A) were outside the CV and were driven with DC 125 V. The power to control the opening and closing of the valves was DC 125 V for all four. To repeat the IC on and off, the normal procedure was to only operate valve 3A while the other three were kept open.

As described earlier, the control power and driving power to the valves were lost at the same time; thus, although the closure signals were sent out, whether the valves closed or not remains unclear. For the following three reasons, however, we believe the IC was barely functioning.

TEPCO's check of the actual IC isolation valves

(1) On April 1, 2011, TEPCO investigated the IC MO valve circuits. If the fail-safe function "failed" to work and the valves were left as the operator intended, *valve 3B should be at "closed" and the others at "open."* The operators did not turn the valves to the "closed" side on or after March 11. The investigation revealed that the electrical circuits of 1A and 4A of System A and 1B and 4B of System B were all indicating "partially open." These indications were probably caused by the fail-safe function sending out "close" signals, but with the driving power to realize the "close" operation lost at the same time, the closing motions had stopped midway.

(2) TEPCO further, on October 18, 2011, checked the actual openings of the four valves outside the CV. The investigation found that valve 2B was closed, verifying that the fail-safe function worked. Reports say that after the SBO, valves 2A and 3A, which the operator hit open after the fail-safe function, were actually confirmed open.

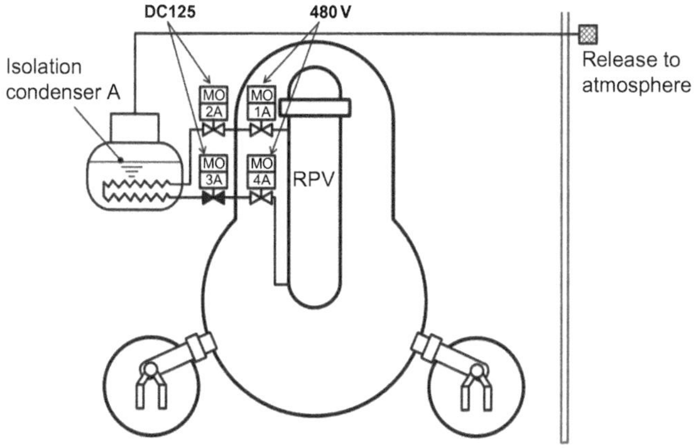

Figure 2.24 IC valve arrangement.

Cooling water left in the condensate tank

On October 18, TEPCO checked the remaining water in the IC condensate storage tank. The tank for System A was 65% full and that for B at 85%. Normally, the amount of cooling water to store is about 80%, and compared to that level, System A was 15% less and System B at the initial state. The IC operation duration to consume 15% of cooling water is, from calculation, about 1 h, and this number backs the assumption that System B was not used, and System A ran intermittently for about 50 min before the tsunami attack and later hardly functioned at all.

Loss of natural convection due to hydrogen

The IC takes in high-temperature steam, cools and condenses it in the condenser, and recycles the water back into the RPV. The driving force, similar to convection, is the difference of specific weights of steam and water. When hydrogen gas lighter than steam mixes in the system, the lightweighted hydrogen stagnates in the upper part of the piping to stop the circulation. IC is a cooling system; however, it stops functioning when core damage takes place and hydrogen is produced. It has a problem in countering severe accidents.

2.5.4 Delay in venting Unit 1

The reasons for the delay in venting Unit 1 from the morning of March 12th to the afternoon follow:

(1) Training of the vent operation had been made only remotely from the central control room, venting under SBO had never been trained, and there was no manual for it. Thus, the staff had to determine from the beginning which valves could be opened by turning handles, where they were located, and what the routes were to reach the locations.
(2) Compressed air was needed to open A/O vent valves; however, no portable compressors had been prepared (Table 2.3).

Table 2.3 Types of vent valves (for S/C vent)

Type	Driving power	Manual handle	Location	Notes
M/O valve	AC 480 V	None	2 F	
A/O valve large	Air	None	Basement 1	
A/O valve small	Air	Unit 1 only	(torus room)	Bypass valve
Rupture disk	None	None	2 F	Rupture pressure 0.448 MPa (gauge)

Note: Rupture pressure for Units 2 and 3 were 0.427 MPa (gauge).

(3) Supplying air to the A/O valves required AC power for opening the air piping valves; however, the AC power had been lost. The workers had to carry in the emergency AC power source originally prepared for temporary lighting.
(4) The workers needed to enter the reactor building to hook up the air; however, high radiation in the building blocked their access.

Venting is an emergency procedure to protect the CV from damage. Historically, the Nuclear Regulatory Commission (NRC) decided to equip reactors with vents in 1989 following remarks that said "Mark-I CV volumes are too small" [4] after the Three Mile Island accident. The purpose is to protect the last boundary, CV, from damage with the sacrifice of releasing some amount of radioactivity.

For venting Unit 1, whether from the S/C route or D/W route, the following three conditions had to be met (Figure 2.25).

(1) A/O valve large or small (bypass valve) is open
(2) M/O valve is open
(3) CV pressure pushed the rupture disk from the inside at pressure 0.448 MPa (gauge) or larger.

The large and small A/O valves were both in the torus room on basement 1. The small one in Unit 1 had a handle for turning it, but the large one did not. (No handles are installed on either the large or small valve with Units 2 and 3.) When the workers started the venting process with Unit 1 after 09:00 on March 12, the radiation level in basement 1 was too high, and they could not enter it to open the small A/O valve. They had to connect a compressor to send air from a location with relatively low radiation levels. The solenoid on A/O further needed AC power. The M/O valves, on

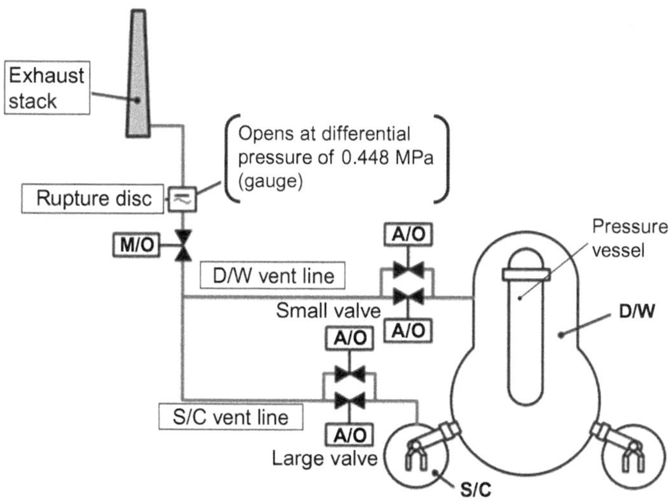

Figure 2.25 Vent lines. There were two lines, one from S/C and the other from D/W. Three closures, at least one of the A/O valves of the two lines, the M/O valve, and the rupture disc, had to be open for the vent.

the second floor of the reactor building, had handles for manual turning, and in fact, the operator on-duty on March 12th opened them after 09:00.

Remarks

Torus room: The room in basement 1 of the reactor building where the S/C is located.

To open two vent valves under SBO, the workers needed both a portable AC power and an air compressor. The fact that none were stored in Fukushima-1 indicates poor AM planning. Section 2.10 will return to this point later.

The rupture disk is a type of safety valve that ruptures at 0.448 MPa (in case of Unit 1) or higher. It is a type of fail-safe function to prevent venting when it is not really needed. It is designed to break before the CV pressure reaches the design strength. When both the A/O and M/O valves are open, the rupture disk breaks before the CV pressure reaches its design value to vent the CV. This, however, means that the CV cannot make the first vent at lower pressure. Injecting water into the RPV from an alternative source of D/DFP (discharge pressure about 0.4 MPa) would require lowering the RPV pressure using SRVs, and this requires lowering the CV pressure by venting it. In some situations, there might not be enough slack time to wait for the CV pressure to reach the rupture disk set point. In such cases, the fail-safe function works in the other dangerous direction. "What is safe" depends on the situation. This issue was similar to the fail-safe function of the IC isolation valves.

2.6 The condition of Unit 3 after SBO

2.6.1 *Up to 02:42, March 13, when an operator manually stopped the HPCI*

March 11, about 15:38: Unit 3, SBO

All AC power was lost due to water submergence caused by the tsunami. DC power survived. The central control room shared with Unit 4 was pitch dark. Unit 3 had its DC power switchboard and batteries on the concourse floor (between the first floor and basement); thus, this equipment did not receive water damage and allowed the operators to monitor the RPV pressure, water level, and other crucial parameters. RCIC had started at 15:25 with the "reactor water level high" signal at 15:25 before the loss of AC power.

March 11, 16:03: RCIC manually started

At about this time, the operator on duty manually started the RCIC (remotely). Because DC power had survived in Unit 3, the central control room was able to operate the RCIC and other equipment as well as read parameters off the indicators.

The RCIC of Units 2 and 3 each had a rated capacity of injecting 97 ton of water per hour. The capacity of the condensate storage tank was about 2500 ton; thus, running the RCIC at full capacity would drain the condensate storage tank in about a day.

Water in the condensate storage tank had other tasks and could not spare all its water for the RCIC. The operator had to control the RCIC so it could last through the long hours to come.

Remarks

The RCIC had another source of water, the S/C (about 3000 tons).

March, 12, 11:36: RCIC automatically stopped

The RCIC stopped for some reason, still unknown as of today. The operators looked for the cause and tried to restart it from the central control room, but did not succeed.

March 12, 12:35: HPCI automatically started

The HPCI automatically started on detecting the lowered water level in the reactor (HPCI, like RCIC, used the condensate storage tank or S/C as its source of water). HPCI is also driven by steam from the reactor and thus does not require AC power for running its pump. HPCI is the last resort of emergency core cooling that can inject water into the RPV, even at high pressure. Its rated capacity is 965 ton/h. This high performance quickly raises the reactor water level and stops automatically. Such repetition of ON and OFF will quickly exhaust the battery. The operator, worried about draining the battery, built a bypass route so part of the water discharged from HPCI would run back to the condensate storage tank to control the amount of water injected to the RPV.

The HPCI water injected effectively cooled the reactor, and its pressure gradually dropped. This pressure decrease lowered the HPCI pump discharge pressure. At this time, however, the reactor water level had not been read, and the effect of water injection was not known. This uncertainty caused concerns in the operator's mind about malfunction of the HPCI, which was forced to run in a state different from normal operation.

March 12, 17:30: early preparation of vent line ordered

The D/W pressure at this time was at about 0.3 MPa (abs), about twice the normal pressure. Plant Manager Yoshida, with the experience of taking long hours to vent Unit 1, felt the need to ready the vent line quickly before the radiation level rose in the reactor building and gave orders to hurry in preparing for the vent.

By that time, the plant emergency headquarters, having learned from the experience with Unit 1, were looking into the procedure for venting Units 2 and 3. The investigation revealed that the two units, unlike Unit 1, did not have handles on the large and small A/O valves for venting the RPV. In other words, to perform the vent, compressed air was absolutely necessary. The M/O valves, on the other hand, had handles for manual operation. It seems it is a common practice for nuclear power plants to put manual handles on M/O valves.

2.6.2 Further progress

March 13, 02:42: HPCI manually stopped

At 02:42, the operator manually stopped the HPCI. As described earlier, the operator on duty had concerns about the status of HPCI and, worried if he should stop it, had consulted with the plant response headquarters. Some of the engineers in the headquarters discussed the matter and reached the conclusion that stopping the HPCI could not be helped for the following reasons.

(1) Running HPCI at a rotation slower than its design speed may lead to damage to the HPCI.
(2) Opening the SRVs would lower the reactor pressure to allow low-pressure water injection using the alternative route from D/DFP.

The team relayed their conclusion to the operator; however, this discussion stayed within the team with the fact that HPCI was stopped. The information of HPCI stopping was not shared with the team leader or Plant Manager Yoshida. As we will discuss later, this process may have affected the later progression of events. When the operator was about to manually turn the HPCI off, the SRV indicator was illuminated green, indicating the SRVs were stopped. The fact that the lamp was on meant that the battery was still alive. The operator judged that he could operate the SRVs from the central control room. After stopping the HPCI, however, when the operator tried to activate the SRVs twice at 02:45 and 02:55, they would not open. It is probable that the battery capacity had dropped to a level to illuminate the status indicator lamps, but below a level to "open" the SRVs, a process that requires a large amount of current. This conjecture is based on the success of their openings, 7 h later, when the batteries had recovered.

March 13, about 03:00: alternative water injection from D/DFP was unsuccessful

The HPCI had successfully cooled the reactor to drop its pressure to 0.58 MPa. After stopping, however, the pressure quickly rose to 0.77 MPa at 03:00 on March 13th, and to 4.1 MPa at 03:44 (Figure 2.26). The D/DFP, which is a low-pressure system, did not have enough discharge pressure to inject water into the reactor vessel at a higher pressure. The situation was critical because they depended on water injection from D/DFP and stopped the HPCI, but the reactor pressure increased to a high level that did not allow water injection from D/DFP. The operators that lost methods of injecting water tried to restart HPCI or RCIC in vain. (The failure to restart the HPCI was probably also due to the low battery.)

Plant Manager Yoshida and other main staff at the plant response headquarters learned about this HPCI stoppage shortly before 04:00, more than an hour after it took place. Yoshida believed before hearing this report that alternative water injection from D/DFP was unreliable because even its closest water source of "filtered water tank" for Unit 1 was located over 500 m away in a straight line, and the piping could have been damaged by the earthquake (the piping seismic classification was level C), and the discharge pressure of about 0.4 MPa was lower than that of the fire engines. Yoshida had a

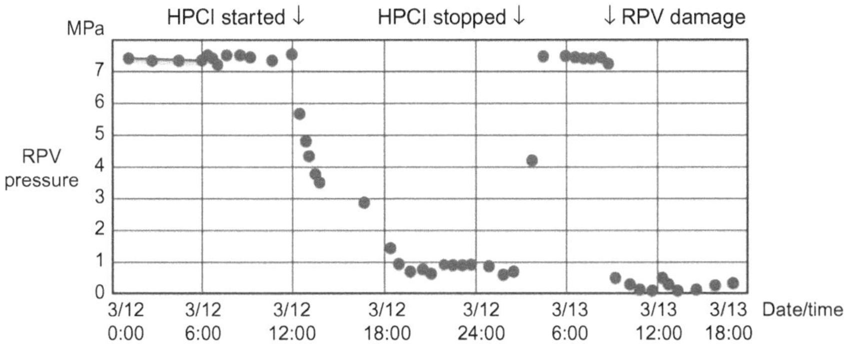

Figure 2.26 RPV pressure from March 12th to 13th. RPV pressure suddenly dropped when HPCI kicked in, and when the HPCI stopped quickly it rose again. The RPV thus, was probably damaged on March 13th.

chance of turning down the idea of manually stopping HPCI if he had been consulted about the process by those on-duty before they actually carried out the process.

Shutting down the HPCI should have been done after verifying an alternative method for water injection was available (i.e., opening the SRVs and testing water injection from D/DFP).

March 13, 04:50: serious preparation for venting

At about this time, the rising reactor pressure without means for water injection forced workers to energize the solenoid valves to open the A/O valves for venting. The operators brought in a portable generator for temporary lighting into the central control room and hooked the cables to the solenoid valve terminals trying to activate the valves. Later, however, when an operator went into the basement of the reactor building to check the A/O valve status, it was closed. The operator that entered the torus room heard the sound of SRV discharging steam into the S/C. When he put his foot on the S/C, it was so hot it partly melted the sole of his shoe.

The operator then, at around 05:23, changed the air bottle to one with air remaining in it and carried out the work to open the large A/O valve. After this work, the recovery team from the plant response headquarters headed to the basement of the reactor building to check the opening of the A/O valve. However, by this time, the radiation level in the area was too high to allow entry, and they had to go back to the central control room. They were still unsure if the A/O valve had actually opened or not.

March 13, about 05:00: information from water level indicator became available

At about this time, for an unknown reason, the reactor water level data became available. The water level indicator showed −2300 mm from the TAF at about 05:10 and −2600 at about 06:00. From the dropping water level, the plant response headquarters assumed it had dropped to TAF at about 04:15. The water level indication at this time was fairly reliable.

At around 07:39, D/W spray (an emergency cooling system that cools inside the D/W with water spray) was activated with water from D/DFP to lower the D/W pressure. At this time the D/DFP system was available.

March 13, about 06:00: preparation of seawater injection from fire engines

From the previous night of March 12th to about this time, seawater injection to Unit 1 was made by connecting three fire engines in series and pumping seawater up from the Unit 3 reverse flow cleaning valve pit. The sudden stoppage of the Unit 3 HPCI forced the workers to start alternative water injection to Unit 3 as well. After 06:00, the other fire engine on standby moved out to form a seawater injection line for Unit 3. The preparation was complete around 07:00.

About this time, however, on the fifth floor of the prime minister's office, Minister Kaieda, Chief Madarame, and others were discussing matters like "Injecting seawater will result in decommissioning the reactor" or "Maybe there is freshwater remaining somewhere" and they decided to inquire if there was freshwater anywhere in the plant. A department manager of TEPCO informed Plant Manager Yoshida about the discussion, and Yoshida acknowledged it as a "strong opinion from the prime minister's office to inject freshwater." He then reconfigured the water injection line from the completed seawater line to a freshwater line from the fire reservoir (Figure 2.27).

The freshwater injection started almost 7 h after stopping the HPCI at about 09:25. The water in the fire reservoir only lasted for about 3 h, and the injection stopped at

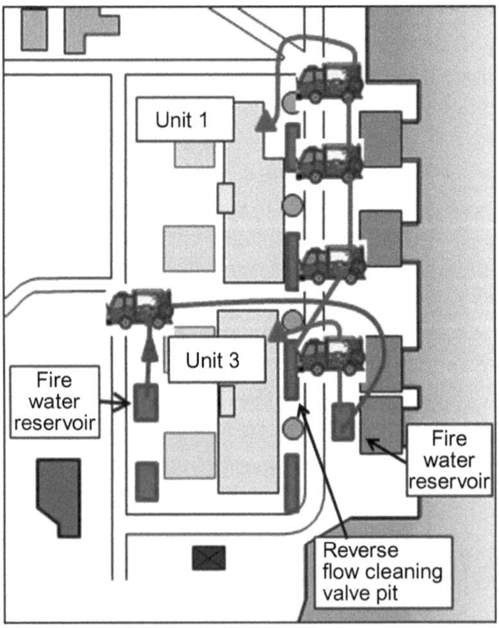

Figure 2.27 Freshwater injection into Unit 3. Water drawn from the fire water reservoir on the west side of the reactor building was injected into Unit 3 via the fire water reservoir on the east side.

about 12:20. Then the workers had to reconfigure the water injection system back to seawater injection, and it had to wait until about 13:12 to start. When we look at the loss of water injection, the freshwater injection not starting until 09:25 was not due to switching the water source, but it was the time lost in waiting for the reactor pressure to drop to the level to allow the injection. The 52-min loss, however, for switching back to seawater after exhausting all the freshwater was due to the change made earlier to freshwater. This loss was an example of directives from Tokyo delaying the team in the field.

March 13, 07:44: securing 12 V batteries

The batteries delivered to the plant from outside up to this point were all 2 V batteries, which were inadequate for activating the SRVs that required 120 V. The plant response headquarters started to search for 12 V batteries and by 07:44 secured 10 batteries from the response headquarters employee cars parked at the plant. These batteries were carried into the central control room in preparation for SRV opening. At around 9:50, over 7 h from stopping the HPCI, the SRVs were opened to release pressure; however, as we will explain later, the pressure vessel had already been depressurized before that from damage to it.

March 13, 08:41: vent line complete

At 08:35, an operator on duty entered the second floor of the reactor building to turn the handle on the vent M/O valve to 15% open. The other A/O valve had already been opened about 4 h earlier. The state was unconfirmed because they could not enter the building. However, at 08:41, the plant response headquarters judged that the vent line was ready (i.e., automatic venting would take place when the CV pressure rises above the rupture disk set point). The headquarters reported so to the ministries and agencies.

March 13, about 09:00: possible damage to the RPV

At about 09:00 on March 13th, the reactor water level shot up, reactor pressure dropped almost instantly, and the D/W pressure rose rapidly (Figure 2.28). The D/W pressure then indicated values that were about the same with the RPV pressure or higher at 0.647 MPa. These data suggest the possibility that part of the fuel dropped to the bottom of the RPV and upon contacting the cooling water at the bottom, generated a large amount of steam to boost the pressure. The RPV bottom was damaged to cause a rapid pressure drop (RPV melt-through). The water level data that followed recorded an abrupt fluctuation. At this time the water level indication may have been out of order, but at the same time, could be showing the real transition of water level jumping up and down with the melted fuel dropping to the RPV bottom and contacting the water to produce an exploding amount of steam.

The Interim Investigation Report (December 2011) states that this rapid pressure drop was from opening the SRVs; however, from the hearings that followed, the committee found it was not. The Final Report corrected the error, and SRV depressurization only took place at about 09:50.

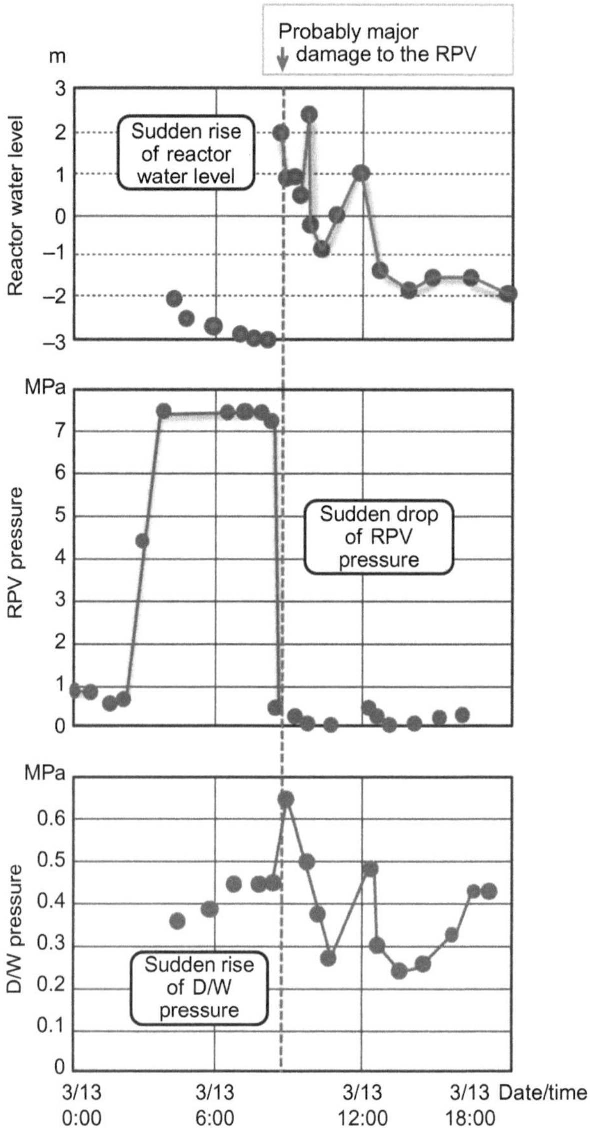

Figure 2.28 Data that hint at RPV melt-through around 09:00 on March 13th.

March 13, 09:20: the vent

The D/W pressure around 09:10 showed a peak value of 0.637 MPa (gauge) above the rupture disk set point of 0.427 MPa (gauge), but 14 min later it dropped to 0.540 MPa. The plant response headquarters determined the S/C was vented at around 09:20 and reported so to the ministries and agents. The two vent valves had been open since about an hour earlier at 08:41, and the venting system was ready to vent automatically

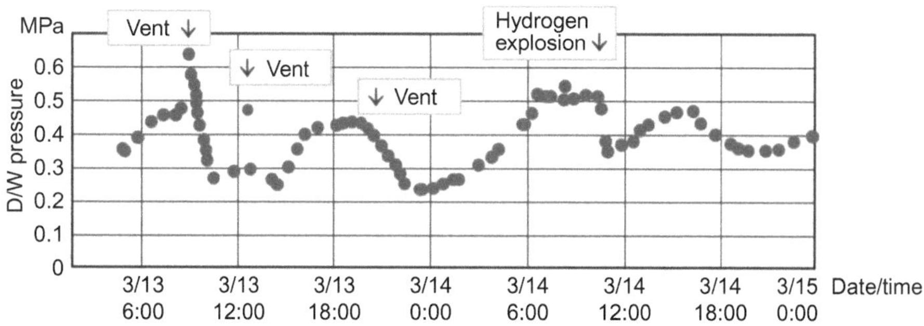

Figure 2.29 Unit 3 D/W pressure transition (13-15th). Could not hold the vent valve "open" due to lack of air. D/W pressure repeated a number of ups and downs.

as soon as the pressure rose above the rupture disk set point. Further at about 10:55, the D/W pressure dropped to 0.27 MPa (Figure 2.29). This suggests a large amount of gas exhaust with radioactive material was made around this time from the exhaust stack to the atmosphere.

March 13, 11:17: D/W pressure rose again, difficulty with keeping the vent valves "open"

Around this time, the D/W pressure started to rise again. The plant response headquarters estimated that pressure drop in the air bottle caused the opening of the vent valves to narrow. Workers entered the reactor building with high radiation levels in two shifts to take turns in 15-min intervals to exchange the bottles. At 12:30, the large A/O valve was confirmed "open," and the D/W pressure, which was at 0.48 MPa at 12:40 dropped to 0.23 MPa at 14:30.

The D/W pressure, however, soon turned around at about 15:00 and went up to reach 0.41 MPa at about 16:45. The plant response headquarters decided there was a limit to being able to keep changing the bottles by entering the building with high radiation and started to discuss a new method using a compressor. The headquarters then managed to borrow a portable compressor from a contractor and started it at around 19:00. The compressor, however, was a small one, and the D/W pressure was still at 0.425 MPa at 20:30. After about 2 h at around 22:30, probably because all the piping was filled with air, the D/W pressure started on its way down to 0.24 MPa at 0:00 on March 14th. The headquarters for the first time managed to keep the large A/O valve open. So from the time they first opened the vent valve at 8:41 on March 13th, they had struggled with it for over 15 h.

Now, let's roll back the time by about half a day.

March 13, 13:12: seawater injection started again

As we showed earlier, freshwater in the fire reservoir was exhausted, and after 52 min of interruption, the water source was switched to seawater in the reverse flow cleaning valve pit, and the water injection started again.

March 13, about 14:31: a number of dangerous symptoms

At about this time, a high level of 300 mSv/h was measured at the north side of the double door, and a white haze was seen inside the door. The workers feared a hydrogen explosion like that in Unit 1, but there was no effective way of stopping it. They even considered shooting a hole open on the reactor building wall with a fighter jet from the self-defense force. They had to abandon the idea because of the possibility of igniting the hydrogen and finally settled on using a water jet. They later started to prepare the tools; however, before they were ready, the hydrogen exploded the next day.

The radiation level in the central control room went up to 12 mSv/h in the Unit 3 side, and the operators had to take cover in the Unit 4 side.

March 14, 01:10: seawater in the reverse flow cleaning valve pit was exhausted, and water injection from fire engine stopped

At about 01:10, the seawater in the reverse flow cleaning valve pit was exhausted, and the fire engine that was injecting water into Unit 3 could no longer draw seawater. The plant response headquarters had been searching for a new water source since the day before (e.g., the water intake near Unit 4, maintenance access hatch over the water discharge channel, directly pumping water up using the fire engine, the basement of the turbine building, and many other possibilities); however, they all were unfeasible. When the fire engine stopped drawing water, another observation of the reverse flow cleaning valve pit revealed there was still some water left and, from about 03:20 on March 14th, managed to restart the water injection. That was probably just a short-term solution, and the headquarters, instead of relying on seawater left in the pit, started to discuss directly pumping water up from the ocean. The primary buildings were standing on land 10 m above sea level; thus, pumping the water with one fire engine would not work, but if they connected two in series, the seawater could be pumped. The workers started building the line, and after 09:00, they completed a line that continuously pumped seawater from the ocean into the reverse flow cleaning valve pit.

March 14, about 03:40: energizing the small A/O vent valve

As Figure 2.29 shows, the vent valve that seemed stabilized went unstable, and at about 03:40 the D/W pressure started again on its way up. Headquarters decided to open the bypass small valve in addition to the main large A/O valve. The operators used the portable small generator for temporary lighting to energize the solenoid of the instrument air (IA) system. Then at about 05:00, they switched the compressor to the portable one delivered from Fukushima-2. The D/W pressure, nonetheless, kept rising until about 07:00 and peaked at 0.52 MPa. Thus, all these efforts did not have as much effect as hoped for.

The D/W pressure transition on March 13th and 14th shown in Figure 2.29 indicate how the D/W pressure continued going up and down because the A/O valve opening was not stable.

March 14, 06:30: workers took cover

Around this time, the D/W pressure was staggered at above 0.5 MPa in a manner similar to Unit 1 before its hydrogen explosion, and workers were concerned with hydrogen leakage from the CV and hydrogen explosion to follow. Plant Manager Yoshida was concerned with the workers' safety,and after consulting with the main office, he ordered temporary retraction to the seismic isolated building around 06:30-06:45. The D/W pressure, however, did not rise any higher, and because building the line to pump seawater from the ocean had priority, Yoshida lifted the order after 07:30.

March 14, 10:53: self-defense force water truck arrived

Around this time, the self-defense force arrived with seven water trucks carrying 35 tons of freshwater. They were headed toward the reverse flow cleaning valve pit in front of the turbine building of Unit 3. When two of the trucks approached the pit, Unit 3 had its hydrogen explosion. Four men on the self-defense force were injured, and the water truck suffered damages. No water was supplied from this truck then.

March 14, 11:01: Unit 3 hydrogen explosion

After a little over 3 h from the lifting of the retraction order, at 11:01, Unit 3 had the hydrogen explosion that everyone was concerned about (Figure 2.30). This explosion injured four personnel from the self-defense force, four TEPCO employees, and three employees of a subcontractor (Nanmei). The workers in the field all took cover in the seismic isolated building. Rumors of a large number of employees missing created a heavy mood in the plant response headquarters. The four fire engines parked in front

Figure 2.30 Ten seconds after hydrogen explosion of Unit 3. The explosion went straight up. Photos © Fukushima Central Television.

of the turbine building all stopped working, and the fire hoses were damaged and unusable. The area was covered with rubbish from the explosion.

Reactor parameters of Unit 3, however, were still available after the explosion. The D/W pressure was at 0.48 MPa and S/C at 0.47 MPa; thus, Yoshida judged that it was a hydrogen explosion of the reactor building, similar to the one with Unit 1.

March 14, after 13:00: response restarted

Observation of the situation in the field revealed that many of the fire engines had stopped with their hoses unusable from the damage. The method of injecting water up to the point before the hydrogen explosion was to pump seawater from the ocean to the reverse flow cleaning valve pit using two fire engines in series and with different fire engines to inject the seawater individually to Units 1 to 3. The area around the Unit 3 reverse flow cleaning valve pit was covered with rubble, and it seemed difficult to rebuild the lines. The headquarters decided to directly inject the seawater from the ocean to Units 2 and 3. During this time, with priority to carry out water injection to these two units, water injection to Unit 1 was halted.

March 14, about 16:30: water injection restarted

The water injection line damaged with the hydrogen explosion was recovered and water injection to the Unit 3 reactor restarted.

March 15, 7:55

Steam was observed above the building, and discussion was made if they should first spray water on the Unit 3 fuel pool instead of the spent fuel pool of Unit 4 with the most number of fuel bundles. Later the spray was first sent onto Unit 3 (Section 2.9).

Table 2.4 Summary of damage to Unit 3 reactor

	Confirmed damage to Unit 3
RPV	• High possibility of damage between about 6:30 and about 9:10 on March 13th • High possibility of further damage by 5:00 on March 14th • Later on March 14th, further severe damage is highly probable
CV	• Possibility of damage between 2:42 on March 13th and 2:20 on March 14th cannot be denied • Plenty of evidence that back the possibility of severe damage between about 7:00 to about 21:35 on March 14th • There is a possibility of further severe damage that followed

Damage: damage that impairs the containment function.

Later

RPV pressure and D/W pressure gradually went down. The CV pressure drop was not rapid; thus, CV damage with Unit 3 was not as bad as that with Unit 2.

Summary of damage to the reactor primary equipment

Table 2.4 shows the damage to the core, RPV, and CV of Unit 3, as they were confirmed by the Investigation Committee.

2.7 The condition of Unit 2 after SBO

2.7.1 Up to about 13:00 on March 14 when the RCIC stopped

March 11, 15:41: Unit-2 SBO

Water submergence from the tsunami caused SBO including DC. The central control room was pitch dark, and all the indicators were not functional. The most important water level and pressure indications of the reactor were unavailable as well. Fortunately, the RCIC had been manually (remotely from the central control room) turned on, but its status from that point was unknown.

March 11, 20:49: temporary lighting turned on, status unknown

A temporary light was turned on in the shared central control room for Units 1 and 2. The water level and injection status, however, were still unknown, and the response headquarters, as preparation against worse situations, prepared to request evacuating the area near the plant.

March 11, about 21:15: pessimistic conjecture of the status

The plant response headquarters, not knowing the RCIC status, estimated the water level to drop to TAF at 21:40, core damage to start at 22:20, and RPV damage at about 23:50. Headquarters assumed the worst for not knowing the water level and other parameters (contrary to their assumption, core cooling was in fact maintained).

March 11, 22:00: water level found to be in a stable condition

Water level was found at TAF + 3400 mm. Around this time, the RPV temperature was at near normal operation level, and the water level indication was also reliable. After about an hour and a half, the RPV pressure was at 6.3 MPa (gauge) and D/W pressure at 0.141 MPa (abs), both in normal ranges, and the engineers were relieved for the moment.

March 12, 01:00-02:55: RCIC confirmed to be operational

At past 01:00, an operator headed toward the RCIC room in the basement of the reactor building to check its status. When he opened the door, however, water gushed out and not being able to enter the room, he headed back to the central control room. At past 02:00, the operator went down to the RCIC room again and managed to get to the entrance, but could not check the RCIC status. He then came up with the idea that he could check the RCIC status by comparing the RPV pressure and the RCIC discharge pressure. He then went to the instrumentation racks on the first and second floors of the reactor building and found the RPV pressure at 5.6 MPa (gauge) and RCIC discharge pressure at 6.0 MPa (gauge). That is, he confirmed the RCIC discharge pressure was slightly higher than the RPV pressure, values that show that the RCIC was running normal. The head of operators on duty then judged that RCIC was operational and reported so to the plant response headquarters.

March 12 about 04:30: RCIC water source switched

The condensate storage tank for each unit was located in front of the turbine building. The capacity was 2500 tons for Units 2 and 3. The operator, concerned with the risk in keeping the RCIC running with the condensate tank, switched the water source to the S/C from the condensate storage tank.

Remarks

The RCIC had two water sources, the condensate storage tank and S/C (about 3000 tons for Units 2 and 3). Using S/C as the water source ensured plenty of cooling water; however, without a heat sink to dump the heat, circulating the water from the reactor to the RCIC and back to the reactor would keep the S/C temperature and pressure up.

March 12, 17:30: order to prepare the vent

From the experience of delaying the vent of Unit 1, Yoshida ordered an early vent preparation. Venting required compressed air to open the A/O valve; however, the only portable compressor available was in use for Unit 1. They decided to open the A/O valve (Figure 2.31) with the air from the air bottle on the IA system piping.

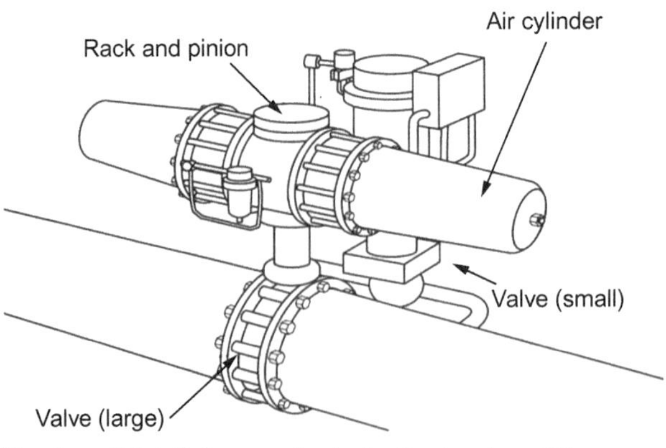

Figure 2.31 A/O valve of Unit 2. Large valve in the front and small bypass valve in the back. Both cannot be opened by hand, and operating them required compressed air.

March 13, 10:15: vent preparation

With the plant manager order at 17:30, an operator went into the reactor building and opened the M/O valve at "25% open." In addition at around this time, the plant manager had ordered to open all valves on the vent line except the rupture disk (i.e., to open the A/O valve as well). The recovery team of the plant response headquarters went into the first floor of the reactor building and opened the discharge valve on the air bottle next to the IA system piping to open the A/O valve. They further used the portable generator for temporary lighting to force the solenoid on the large A/O valve open by energizing it. At about 11:00, they succeeded in opening the large A/O valve to complete the vent line preparation. The S/C pressure at this time, however, was smaller than the rupture disk set pressure of 0.427 MPa (gauge), and the venting did not take place then.

March 13, past 18:00: preparing a compressor

The plant response headquarters decided to stop using air bottles with unknown amount of air left and rely on a portable compressor like Units 1 and 3. Through teleconference, they requested one from other plants, and about 7 h later, one arrived from Fukushima-2 before 02:00 on March 14th (Unit 1 had one since about 12:30 on March 12th and Unit 3 from about 19:00 on March 13th from contractors).

March 13, late in the afternoon: water injection line ready for Unit 2

At about this time, a water injection line was readied for Unit 2 like Units 1 and 3 that used the reverse flow cleaning valve pit as the water source. Unit 2, at this time, however, had its RCIC in operation and the RPV pressure was still a little less than 6 MPa. Water injection with a fire engine was only made ready and kept on standby.

March 14, 11:01: Unit 3 hydrogen explosion, vent line and water injection lines damaged

Unit 3 had a hydrogen explosion. By that time, the central control room had prepared a temporary circuit to energize the A/O valve solenoid. The hydrogen explosion, however, broke the circuit, and the large A/O valve reclosed. The explosion also damaged a fire engine and hoses, but luckily, the compressor survived the blast.

2.7.2 Further progress

March 14, 13:25: RCIC judged to have lost its function

As described earlier, the reactor water level was going down after around 12:00. The water source for the RCIC had been switched to the S/C pool water from around 04:00 on March 12th, meaning the cooling water had been storing heat for over two days.

Thus, the S/C at about 12:30 was at an abnormal state of high temperature and pressure with temperature at about 149 °C and pressure at 0.49 MPa. Given the situation, Plant Manager Yoshida judged that "the RCIC had stopped at about 13:25."

March 14, 14:43: delay in seawater injection

At about this time, the water injection line damaged by the hydrogen explosion had been recovered early in the morning, thanks to great efforts by the workers. The afterquakes, however, had interrupted further work, and the fire engine started only at about 16:30, and water injection to Unit 3 started then. Unit 2, however, took several hours in operating the SRVs delaying the reactor depressurization, and the state of no water injection continued.

March 14, about 15:00: D/W pressure dropped

The D/W pressure at 0.465 MPa (abs) at about 12:30 started to drop after 13:35, and at about 16:57, it had clearly dropped to 0.40 MPa (Figure 2.32). During this time, no venting or CV cooling system activation had taken place. The D/W pressure drop was, thus, probably due to leakage from the CV that had been subject to high temperature and pressure for an extended time.

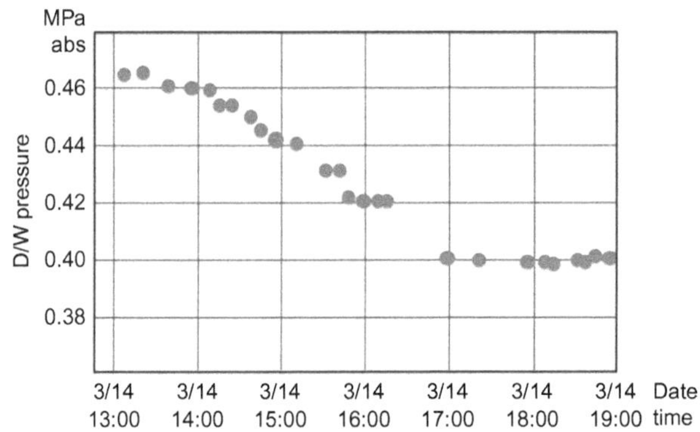

Figure 2.32 D/W pressure drop on March 14th. Gradual D/W pressure drop likely to be caused by leakage from CV.

March 14, about 16:00: vent valve did not open

The recovery team of the plant response headquarters kept energizing the solenoid from the central control room; however, as mentioned earlier, the team could not keep the A/O valve open. This was possibly due to the small size of the compressor (see Remarks). At 18:35, the operators tried to open the small valve, in addition to the large A/O valve. They succeeded in cracking it open; however, at this time, the CV pressure had not reached the rupture disk set pressure and in any case, the status of no venting continued.

Remarks

Air piping damage due to the earthquake was another possible reason that cannot be denied.

March 14, about 16:34: SRVs did not open

At about this time, the operators in the central control room connected the ten 12 V batteries collected from around the plant in series and tried to activate the solenoids of the SRVs; however, the SRVs did not open.

March 14, 18:22: all fuel bundles exposed

At 18:22, the water level gauge indicated TAF −3700 mm, and at 18:50 it went off scale at the lower end, and the water level could no longer be measured. The plant and main office response headquarters both recognized that all fuel bundles were exposed from water at about 18:22.

March 14, 19:03: RPV depressurization with SRVs

At about this time, the SRVs finally opened, and the reactor pressure went down to 0.63 MPa to allow water injection. At about this time, Unit 3 had seawater injection drawn directly from the ocean and relayed with two fire engines connected in series with hoses. Unit-2 was ready for water injection in a similar manner without having to pour the seawater into the reverse flow cleaning valve pit.

March 14, about 19:20: the fire engines ran out of gas

At about this time, water injection to Unit 2 had just started; however, the fire engines were found to have run out of gas. TEPCO's self-fire defense team, in response, supplied gas from their gas tank trucks; however, water injection to Units 2 and 3 was suspended for at least 37 min.

March 14, 19:57: continuous water injection to Unit-2 started but unstable

Unit 2 without any water injection after its RCIC stoppage, finally received continuous water injection from a fire engine.

Although the water injection had restarted, the Unit 2 RPV pressure was most of the time (over half of the time) from about 21:00 on March 14th to about 01:00 the next day, at a level higher than 1 MPa. Thus, during this time, water injection from the fire engine was physically impossible. Damage to the core and RPV progressed then, and the D/W pressure probably went up and stayed there.

March 14, about 23:35: decision made to vent the D/W

The biggest problem with Unit 2 at this time was that it had not succeeded in depressurizing the CV with S/C venting. The RPV had the safety valve SRVs, whereas the CV did not have equivalent equipment. If the vent continued to be unavailable, the "last resort" CV had the danger of exploding.

At about this time, the S/C pressure was showing values lower than the rupture disk set point, and the D/W, on the other hand, was showing higher values of 0.7 MPas above the rupture set point (see Remarks). The staff at the response headquarters, including the main office, reluctantly decided to vent the D/W. As mentioned earlier, D/W venting was different from S/C venting because it would release the gas inside the CV directly into the atmosphere without passing it through water and thus would cause radiation release that was more than one hundred times that from S/C venting. It was a process to avoid at all costs; however, it was judged that protecting the CV from destruction had higher priority.

Remarks

As we will discuss later, the S/C pressure indicator was likely to have been showing incorrect values.

March 15 about 00:16-01:11: highly dangerous state with high CV pressure

As Figure 2.33 shows, the RPV pressure exceeded 1 MPa three times between 20:30 on March 14th and 01:30 on March 15th. These large fluctuations can be explained as follows: With the RPV already damaged and leaking, the melted fuel and water came into contact to cause strong evaporation of the water to steam that rapidly increased the

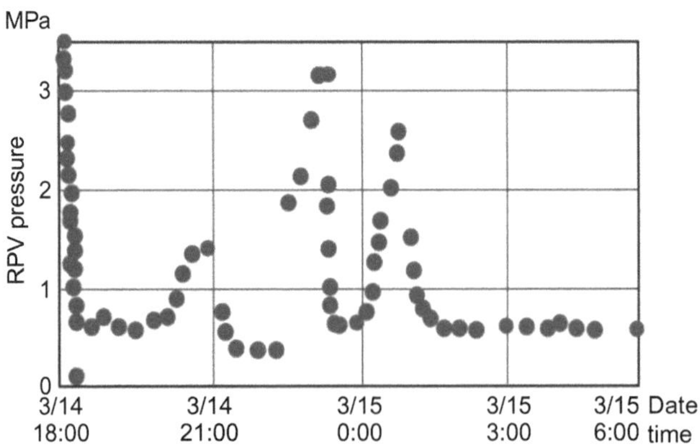

Figure 2.33 RPV pressure transition on 14th and 15th, an indication of possible contact of melted fuel and water.

RPV pressure, and then with leakage from the RPV, the pressure dropped again. The phenomena repeated to produce the cycle in Figure 2.33.

Next as Figure 2.32 shows, the D/W pressure went down to about 0.4 MPa from about 14:00 to about 17:00 on March 14th. In general, when a vessel has leakage, its pressure will keep going down; however, as Figure 2.34 shows, the D/W pressure quickly went up around 23:00. The reason for this transition is as follows: The CV had already started its leakage, and after about 20:00, large leakage from the RPV started as well. Thus the CV was leaking to the outside but at the same time had RPV leaking into it. The rapid rise in the D/W pressure around 23:00 was caused by stronger leakage into the D/W from the RPV.

The radiation (γ-ray) monitoring results from CAMS (Figure 2.35) also support this conjecture. At 15:15 on March 14th, S/C was at about 0.01 Sv/h and D/W its tenth at about 0.002 Sv/h. After 6 h and 40 min of interruption, the restarted measurement at

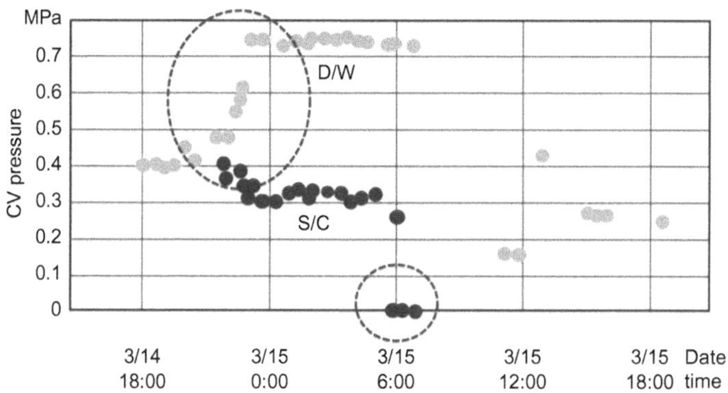

Figure 2.34 CV pressure transition on 14th and 15th. D/W shows a pressure increase despite leakage to the outside, and S/C, which should be at about the same pressure, shows a sudden drop to near zero (instrumentation error).

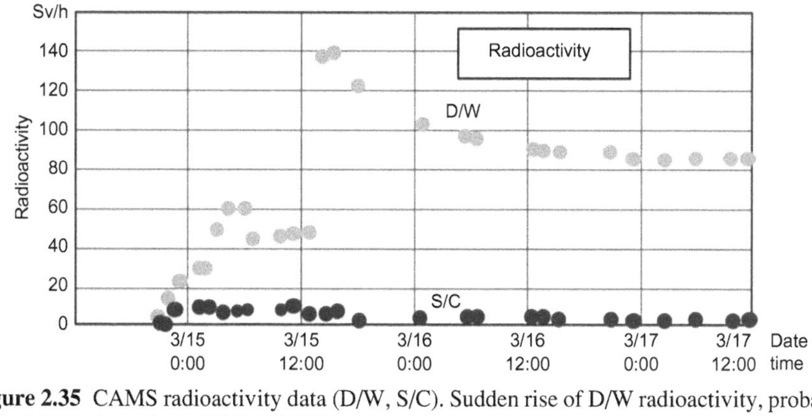

Figure 2.35 CAMS radioactivity data (D/W, S/C). Sudden rise of D/W radioactivity, probably caused by leakage from RPV to D/W.

21:55 showed that the D/W radiation had *rapidly risen to exceed that of the S/C* (Figure 2.34). This was probably due to leakage from the RPV directly to the D/W. Thus, the rapid and large D/W pressure rise starting at about 23:00 in Figure 2.34 matches the data from CAMS.

Summarizing the situations around this time, water injection from the fire engine to the RPV with pressure above 1 MPa was impossible. The CV venting failed, and the D/W pressure rapidly went up around 23:00 on the 14th to reach 0.7 MPa in about 30 min and stayed at this dangerous level, which could have resulted in CV explosion. In the midst of such a dangerous situation, Plant Manager Yoshida then consulted with the main office response headquarters and decided to keep, in the worst-case situation, only necessary personnel for each plant control and to evacuate the others from the Fukushima-1 site. To avoid a sense of panic to the others, he gave instructions to only a few to prepare buses so workers could flee the site if the situation worsened.

There is no question that both the RPV and CV had suffered damage at midnight on March 14th; however, the staff did not think that leakage from the CV had started. They were concerned with not being able to vent the CV and were worried about the worst-case situation of its explosion. A loud sound was then heard at 06:10.

March 15, 06:00-06:10: loud sound

It was when the shift was about to turn over when they heard a loud explosive sound in and around the central control room. They were all then given instructions to take cover in the seismic isolated building, and when they exited the building, they were surprised to see that the surrounding scenery had drastically changed. This exploding sound, as we will discuss later, was in fact the sound of the hydrogen explosion of Unit 4. The situation that had surrounded the staff at the time of not succeeding with the vent and the D/W pressure stopped at a high value of 0.7 MPa had put the staff in a state where everyone was holding his breath. The big bang shook the building, and many of them feared the worst. Under the situation, the plant response headquarters kept the minimum necessary staff of 50 on site and evacuated the remaining 650 workers temporarily to Fukushima-2.

The S/C pressure immediately after the explosion was pointing at zero, and with the sound they heard, the people thought that the CV had exploded. The time that the S/C pressure dropped to zero was, however, a few minutes before the exploding sound. Also, this S/C pressure indication was in absolute pressure, and an absolute pressure of zero would mean vacuum, and that physically could not have happened. For these two reasons, the judgment that "the CV exploded and its pressure dropped to zero" was wrong. Also, as Figure 2.34 shows, the S/C pressure was pointing at about half the D/W pressure and suddenly dropped to zero at 06:02. A situation that the S/C pressure was half of that of the D/W pressure for an extended time is also physically hard to imagine (see Remarks). From these facts, we can reason that the S/C pressure indication was unreliable and the values were wrong.

Remarks

If the D/W pressure was higher than the S/C pressure, gas in the D/W would travel through the vent pipe to exit into the S/C pool. In the reverse situation, gas in the S/C would pass the vacuum rupture disk to go into the D/W. Thus, usually there is not much of a pressure difference between the two volumes.

After about an hour, at 07:20, the D/W pressure recorded 0.73 MPa (abs) and then after another long interruption of over 4 h at 11:25, it went down to 0.155 MPa, a value close to atmospheric pressure. It is probable that the CV suffered significant damage during this time.

Later, the monitoring post near the main gate showed extremely high radiation values on March 15th and 16th with a peak at 10,000 μSv/h (around 08:00 on March 15th). The radiation was the highest on March 15th to 16th, and unfortunately, the wind changed into a relatively strong southeast one blowing from the ocean toward the land (Figure 2.36). Also at night, 1.5 mm of cold rain fell for over an hour. This wind and rain dropped a large amount of radiation over the area northwest of the plant and caused serious radiation contamination. These were unfortunate weather conditions for the local districts of Iitate, Namie, and others northeast of the plant.

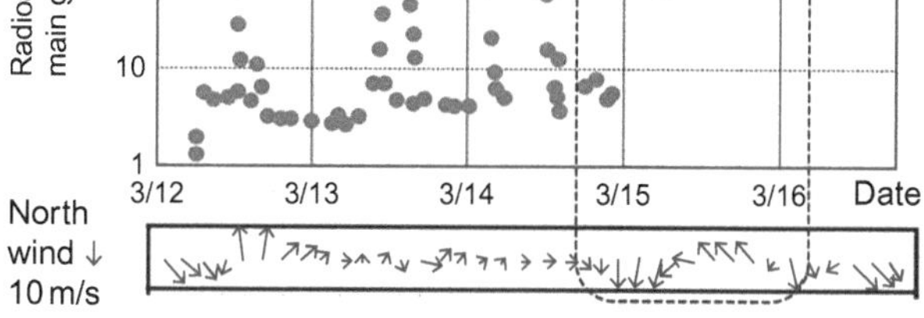

Figure 2.36 Radiation dose and wind near main gate during March 12th–16th. *Source:* NISA, December 27, 2011, "Discussion on containing function".

March 15, 11:25: lifting the temporary evacuation order

As mentioned earlier, the high D/W pressure of above 0.7 MPa was found to have dropped to 0.155 MPa. Those that had evacuated Fukushima-1 to take cover in Fukushima-2 gradually returned to the plant response headquarters starting with group managers.

Later

The RPV and D/W pressure values remained at about the same level with the atmosphere. Radiation reading from CAMS for the D/W did not drop for a long time.

Summary of damage to the reactor primary equipment

Table 2.5 shows the damage to the core, RPV, and CV of Unit 2, as they were confirmed by the Investigation Committee.

Table 2.5 Summary of damage to Unit 2 reactor

	Confirmed damage to Unit 2
RPV	• Damage is likely to have happened by 21:18 on March 14th • High possibility of severe damage later
CV	• Plenty of evidence that damage occurred between about 13:45 and 18:10 on March 14th • Possibility of further damage is very high • High possibility of severe damage to have released a large amount of radiation from 7:38 on March 15th to about 4:00 on March 16th

Damage: damage that impairs the containment function.

2.7.3 Exploding sound at 06:10 on March 15

Source of the exploding sound

The tremor from the explosion was caught on a number of seismographs, and the time of the sound was 06:12. The difference between the arrival time of the primary wave (P wave) and shear wave (S wave) at multiple locations allow spotting the source. The data identified the source to be around Unit 4 and not Unit 2. The exploding sound heard at 06:12 was, without question, the hydrogen explosion in Unit 4, and because there was no evidence of another explosive sound, there is no reason to believe that the Unit 2 CV exploded.

About "zero S/C pressure"

When the workers heard the exploding sound, the S/C pressure went to zero, and many thought that there was an explosive destruction around the S/C. TEPCO's data, however, recorded S/C pressure at zero twice at 06:02 and 06:10 before the hydrogen explosion in Unit 4 (06:12). And as we mentioned earlier, the S/C pressure indication was the absolute pressure, and there is no explanation to support that the S/C was at vacuum. Also, the D/W pressure around this time was at 0.7 MPa, and from the CV structure, the two pressure values of the D/W and S/C were unlikely to have large differences. The S/C pressure gauge of Unit 2, thus, was likely broken at this time. Data from pressure gauges in general are reliable. This was the only event with a troubled pressure gauge in the Investigation Report.

Therefore, the thoughts of many workers in the field that there had been a big explosion with the CV near the S/C and the S/C pressure drop to zero resulted from the fear of such an event in the people's minds. The hydrogen explosion in Unit 4 happened to take place at about the time of zero S/C pressure indication.

On the other hand, there is hardly any question about a large-scale leakage of radioactivity from Unit 2 from March 15th on, and the denial of an explosion does not mean the damage to the Unit 2 CV was small.

2.8 Sequence of events leading to hydrogen explosions in Units 1, 3, and 4

Three units, 1, 3, and 4, had hydrogen explosions in their reactor buildings one after another. The explosions themselves scattered radioactive material that had leaked into the CV, but did not add to the amount of leakage from the CVs. They did, however, cause much trouble in the recovery work and were among the major factors that magnified the seriousness of the accident.

Hardly any of the world experts had expected that the reactor buildings would have hydrogen explosions. It seems odd for an outsider to know that many papers had been published about hydrogen explosions in the CV, but hardly any on that in the buildings (see Remarks). In an interview, one expert attested that he had the impression that the reactor building was being ventilated. On March 11, when asked by Prime Minister Kan, "Is there a possibility of hydrogen explosion?" NSC Chief Madarame answered, "Do not worry because it is filled with nitrogen," and that was a reply that clearly shows he was only concerned about a hydrogen explosion in the CV.

Remarks

The Investigation Committee conducted a literature search to find only two papers on hydrogen explosion in reactor buildings and could not find a trace of discussion in domestic and international organizations, including the International Atomic Energy Agency (IAEA).

2.8.1 Differences between the explosions in the units

Units 1 and 3 suffered severe damage on the top (fifth) floor, and Unit 4 on the fourth and fifth floors. The difference is probably because the explosion was centered on the fifth floor in case of Units 1 and 3, whereas, for Unit 4 the center of explosion was on the fourth floor. The volume in the reactor building fourth floor is about a fifth of that in the fifth floor; thus, the explosion in Unit 4 was made with a smaller amount of hydrogen compared to Units 1 and 3.

For Units 1 and 3, possible routes of hydrogen leakage were the flange at the CV top, access hatch, equipment hatch, instrumentation wire feed-through, and the air

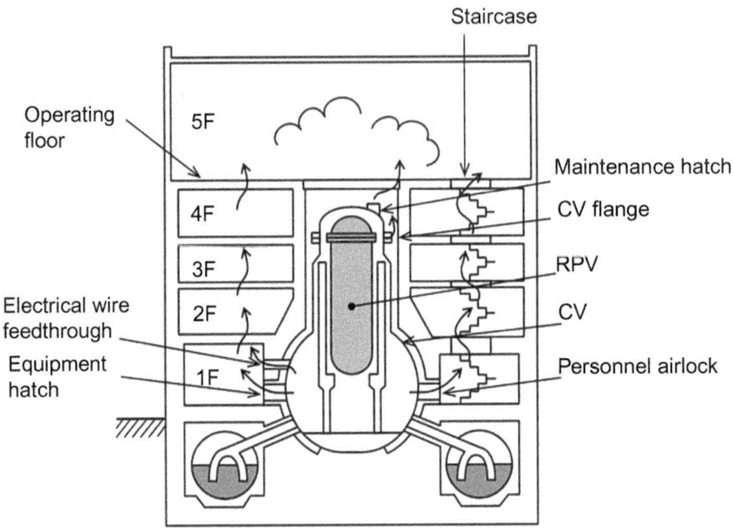

Figure 2.37 Possible hydrogen escape routes.

lock (for worker entry to the CV; Figure 2.37). At the CV flange, the tightening bolts could have elongated due to the pressure rise producing gaps in the bearing surface. Other routes usually have tight seals with O-rings or other packing material like resin, and these polymeric materials were likely to have suffered damage from the high temperatures.

The hydrogen that exploded in Unit 4, which had been shut down for periodic maintenance, must have entered the reactor building of Unit 4 from the exhaust tower shared with Unit 3 when Unit 3 was vented.

Comparing the explosion photographs of Unit 1 and Unit 3, the blast expanded horizontally with Unit 1, whereas that of Unit 3 shot up into the sky (Figure 2.38). This difference is likely to have been caused by the difference in the building structures.

The walls on the fifth floor of Unit 1 were panels attached to steel frames (Figure 2.39); thus, the walls could easily blow out sideways to release the pressure from

Figure 2.38 Difference in the explosions of Unit 1 (left) and Unit 3 (right). Both photographs show 4 s after the explosions. Smoke went up high with Unit 3; however, Unit 1 is estimated to have generated more hydrogen.
Photos © Fukushima Central Television.

Figure 2.39 Unit-1 walls on the fifth floor were panels fixed to the steel beams and thus were weak.
Photo: Air Photo Service.

the explosion. The ceiling did not scatter and simply fell on the fifth floor, while the walls were all blasted away. There was hardly any damage to the fourth floor and lower levels.

Walls on the fifth floor of Unit 3, on the other hand, were reinforced concrete continued from the lower floors and thus were sturdy (Figure 2.40) in contrast to the ceiling with steel frames and panels with Unit 1. The explosion on the fifth floor of Unit 3 did not get the horizontal relief and thus had to shoot up high. The blast took out the ceiling, all walls, and some of the walls on the fourth floor. These left concrete rubble on the ground that hindered the recovery work that followed.

Figure 2.40 Unit 3 on March 16th. The surrounding walls were made of reinforced concrete, and their strength resulted in great explosive power.

It appeared that the Unit 3 explosion had more energy than Unit 1; however, the one with Unit 1 was calculated to have had about 1.5 times the hydrogen with Unit 3. Unit 3 had a stronger fifth floor, and thus more energy accumulated in it before the explosion wiped the floor out.

2.8.2 Investigating whether the explosions were caused by hydrogen produced from core damage

The Investigation Committee Report discusses in detail the argument of whether these explosions were caused by hydrogen from core damage. The following paragraphs discuss the points in summary.

The only explosive material was hydrogen

Both Units 1 and 3 maintained RPV and CV pressure values several times that of atmospheric pressure; thus the possibility of hydrogen explosion with the RPV or CV is denied. If such an explosion took place inside the vessels and blasted the building walls, the vessels would have been destroyed as well, and their pressure values would have been at the atmospheric level.

Also, time had passed since the earthquake and thus there was no chance of dust from the shaking of the earthquake causing a dust explosion. An explosion from combustible gas was the most likely cause, and because there was no combustible gas in an amount to possibly cause an explosion other than hydrogen, it was identified as the cause.

There was enough hydrogen to cause an explosion

To have a hydrogen blast on the fifth floor of Unit 1, assuming the volumetric low limit of hydrogen explosion at 18.3%, about 312 kg of hydrogen was needed. On the other hand, TEPCO analysis announced in March 2012 and the Japan Nuclear Energy Safety Organization (JNES) analysis announced in September of the same year both concluded that about 900 kg of hydrogen had generated in the reactor. Thus the core damage had probably produced enough hydrogen to blow up the fifth floor of Unit 1.

For Unit 3, similar calculation returns 371 kg as the necessary amount of hydrogen. TEPCO's analysis of the generated hydrogen was over 600 kg and that of JNES 550-700 kg. Even if the vent had released some into the air, there was plenty of hydrogen left to explode.

The hydrogen explosion of Unit 4 took place on the fourth floor, which had about a fifth of the volume compared to its fifth floor. The minimum amount of hydrogen necessary to cause an explosion was calculated to be 74.2 kg, and this amount is reasonable as an amount that trickled in from Unit 3 (Figure 2.41).

Reasons why Unit 2 didn't explode

When the fifth floor of the reactor building of the adjacent Unit 1 exploded, the blowout panel of Unit 2 opened like a window, and hydrogen never accumulated in the top floor (Figure 2.42). What seemed like damage from an event turned around and saved the building.

Figure 2.41 Unit 4 on March 23rd. The outside walls were made of reinforced concrete. Half of the walls were gone, leaving the columns and beams.

We often hear that this blowout panel not popping open with Units 1 and 3 was a backfire from a lesson from the trouble at Kashiwazaki-Kariwa NPP during The Chuetsu Offshore Earthquake. During this 2007 earthquake, a blowout panel dropped and created some concerns. Later as a countermeasure, involving other plants as well, the panel was secured tighter to its frame. The Investigation Committee, however, learned that the modification had not been in place yet for Unit 3. The blowout panel, nonetheless, did not pop open even with the high CV pressure.

Figure 2.42 Unit 2. Steam was coming out from the opening after the blowout panel dropped. Photo: Air Photo Service.

After the hydrogen explosion in Unit 1, the headquarters saw that Unit 3 would also have a hydrogen explosion for sure, and discussions were begun to open a hole in the reactor building. The conclusion was to use water jet; however, the plan was not completed in time, and the explosion took place on March 14th.

2.9 The process of water injection into the spent fuel pools

2.9.1 *Plant status up to March 15th*

Early in the morning of March 13th, a stack of white smoke was seen to come up from Unit 1. This caused concerns within the plant and main office response headquarters that if spent fuel was exposed from water in the SFP, the spent fuel without any cover would release a large amount of radioactivity. Unit 4 had the biggest number of fuel bundles in storage, and its SFP was of great concern. Discussions were started about spraying water from a fire engine with an extended ladder or a helicopter, but they could not come up with an effective idea until the hydrogen explosion in Unit 4 at past 06:00 on March 15th.

2.9.2 *March 15, about 09:00: started discussion about water injection into the SFP*

The spent fuel in Unit 1 had passed over 1 year after they were removed from the core, and their small decay heat gave them low priority for water injection. The SFP of Unit 1, thus, had water injection for the first time on March 31 from a concrete pump truck (about 90 ton of freshwater). The following injection was not made until May 20. Unit 2 had its reactor building intact and thus could not receive water spray from above. The workers performed water injection from the fuel pool cooling and cleanup system using a fire engine, and the SFP was filled up full on March 22nd.

On March 15th, the response headquarters had to decide which SFP to put priority on: Unit 3 or Unit 4. Unit 4 with the largest amount of spent fuels and many still with large decay heat first received the priority (Table 2.6); however, in the morning of the next day, white smoke was seen from Unit 3, and they decided to make the judgment after an observation from a helicopter.

2.9.3 *March 16, afternoon: visuals of Unit 4 SFP*

A TEPCO employee boarded a self-defense force helicopter and observed the plant. He confirmed high enough water level with the Unit 4 SFP covering the fuel bundles. The mechanism for this water in the SFP was explained as follows: Unit 4 was in its periodic maintenance, and SFP and the reactor well, on the other side of the pool gate, were filled with water. After the accident when the SFP water level

Table 2.6 **Number of fuel bundles stored in SFP**

	Number of stored fuel bundles		
	Spent	**New**	**Total**
Unit 1	292	100	392
Unit 2	587	28	615
Unit 3	514	52	566
Unit 4	1331	204	1535

lowered with evaporation caused by decay heat, the pressure on the well side, with higher water level, of the pool gate pushed the pool gate to crack open, and water in the reactor well flew into the SFP (Figure 2.43). Given the good condition of the Unit 4 SFP, they decided to prioritize spraying water onto the Unit 3 SFP.

Remarks

If Unit 4 did not have this "luck" and the SFP was left without water injection or spraying until March 20th, what would have happened to the fuel in the SFP is a question for the experts to answer.

2.9.4 March 17: started water spraying

First, for about 13 min from 9:48, two self-defense force helicopters made a total of four trips to dump 30 tons of seawater onto Unit 3. Many who watched this process on TV felt the process was not very effective. Water spray from helicopters was only made this one time. Then for about 8 min from 19:05 on the same day,

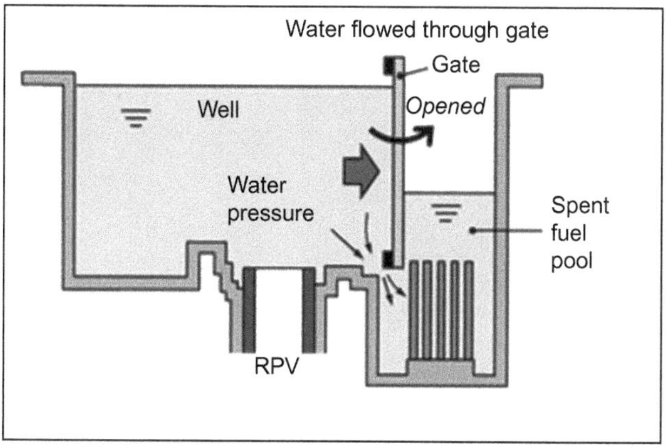

Figure 2.43 Unit 4 SFP. Luckily the gate opened, and water in the adjacent well flowed into the SFP.

the Tokyo Metropolitan Police Armed Response Team sprayed 44 tons of seawater from a high-pressure fire engine. This attempt also lacked effect with the short reach of sprayed water and was only performed this once also. Next a fire engine from the self-defense force sprayed 5 times and a total of 30 tons of freshwater (Figure 2.44).

On the following day, March 18th, the self-defense force fire engine sprayed 40 tons, then TEPCO armed with a U.S. military fire engine sprayed 2 ton of freshwater. On March 19th before dawn, the Tokyo Fire Department sprayed 60 tons of seawater from its fire engine, and the seawater spray on Unit 3 significantly increased from then on.

2.9.5 March 20: water spray to Unit 4

From March 20th, water spray to Unit 4 started. The self-defense force sprayed 80 tons of freshwater twice bringing the total to 160 tons. Most of this water was judged to have landed on the SFP. Spraying water from the ground ended on March 25 with the last spray on Unit 3.

2.9.6 March 22: concrete pump truck

From March 22nd, a concrete pump truck with a special long arm of about 60 m (a.k.a. Giraffe) came onto the scene to spray water onto the target spot (Figure 2.45). It started with Unit 4 on March 22nd and moved to Unit 3 on March 27th. This truck made the water spraying a reliable process, and other fire engines were no longer needed. The Giraffe kept pumping water until June.

Figure 2.44 Unit 4 on March 23rd. Spraying water from the ground had some effect in cooling.

Figure 2.45 Concrete pumping vehicle at work. The concrete pumping vehicle (nicknamed Giraffe) saved the SFP from danger at the time.

2.10 How the accident might have been avoided

2.10.1 Safety measures in place in other countries

A number of safety measures, not taken in Japan, were in place in other countries. The following are six such examples that would prove effective during severe accidents:

(1) Emergency power
Browns Ferry NPP (Mark-I) in the United States has a portable battery rack to serve as an 8 h emergency power supply to take readings from instruments (Figure 2.46) [3].

(2) Watertight doors
This is another preparation at Browns Ferry. The emergency D/Gs are located in a room with watertight doors [4].

(3) Manual handle for IC valves
Milstone NPP (Mark-I, U.S.) has handles on the IC valves inside the CV and conducts training to open the valves from outside the CV [5]. Fukushima-1 Unit 1 IC valves inside the CV could not be opened from the outside.

(4) Snorkel
Diablo Canyon NPP (PWR) has air inlets for seawater pumps at 13.5 m above sea level using snorkels [6] (Figure 2.47).

(5) Vent filter
Mühleberg NPP (Mark-I, Switzerland) has a filtering system on its vent lines that will filter out many of the radioactive materials by passing the vent gas though water mixed with special liquid ingredients [3]. The liquid is injected by gravity so no electricity is needed.

(6) Independent emergency cooling system
The Mühleberg NPP has developed, in addition to the GE-designed cooling system, an entirely independent emergency cooling system for each building [3]. Inside the building is a watertight room with an emergency D/G and a switchboard.

Figure 2.46 Emergency battery power unit. After 9/11, power plants in the United States enhanced their preparation against SBO.

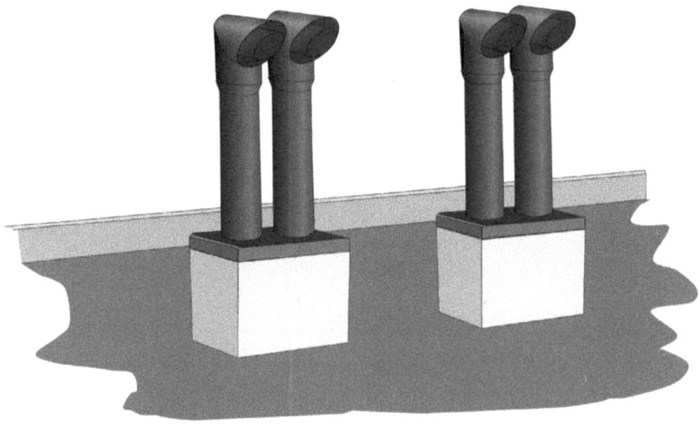

Figure 2.47 Snorkel vent holes. Designed to withstand tsunami waves up to 13.5 m. *Source:* Pacific Gas and Electric.

The Japanese nuclear power generation technologies were superior in material science, machine reliability, and damage prevention in case of an earthquake. Most of these measures are techniques for preventing small accidents, but there was no safety engineering for disaster severity reduction in case a large-scale accident took place. In other words, we can sum up the problem to "We concentrated our efforts to prevent small accidents, but we failed to pay sufficient attention in planning what to do in case a large accident occurred."

2.10.2 Measures that were feasible

It is a misleading mind-set to think building tall seawalls is the only way to prevent tsunami damage. There are a number of safety measures are realistic and not so costly. Instead of planning to prevent even the smallest water submergence, we need to change our mind-set and think to draw a safety line to prevent only the worst accident of core damage. We then realize the importance of protecting peripheral but indispensable equipment for severe accident prevention, like switchboards, batteries, and compressed air are equally important like the main facilities. Ensuring safety to such equipment with back-ups and watertight rooms would probably cost a relatively small amount. We would like to point to the following five measures from the viewpoint of what the minimum measures were that could have been prepared simply and inexpensively to prevent severe accidents (Figure 2.48).

(1) Providing Variety with Switchboard Installation
The lack of variety in the installation of switchboards was a definite factor in failing the deep protection. Multiple switchboards should have been distributed in a variety of locations.
(2) Preparation for Loss of DC Power
Loss of DC power caused difficulties with the control and instrumentation function at the plants. It was a fatal factor in the accident response. The stoppage of the Unit 1 IC with its fail-safe function was also directly caused by the loss of DC power. Although it might be hindsight, the plant should have kept a stock of at least 12 V batteries easily found in shelves of auto shops.
(3) Providing Watertight Structures with Buildings
Retrofitting buildings to make them watertight would not cost too much. Even if modifying the entire building turns out difficult, at least reworking rooms with primary equipment to make them watertight is an inexpensive and feasible measure.
(4) Storing Portable Air Compressors
The air compressor stopped working with this accident, and that was the main reason in delaying the opening of the A/O vent valves. Air compressors are inexpensive machines, and the plant should have stocked portable air compressors with generators in case of AC power loss.
(5) Improving the Water Level Gauge
This accident taught us that when the core damage causes a high temperature in the CV, the water level in the reference condensate pot goes down, and the gauge will no longer be effective. Ideas for improving this instrument will have to wait for further discussion; however, this is clearly a serious weakness for such a crucial instrument.

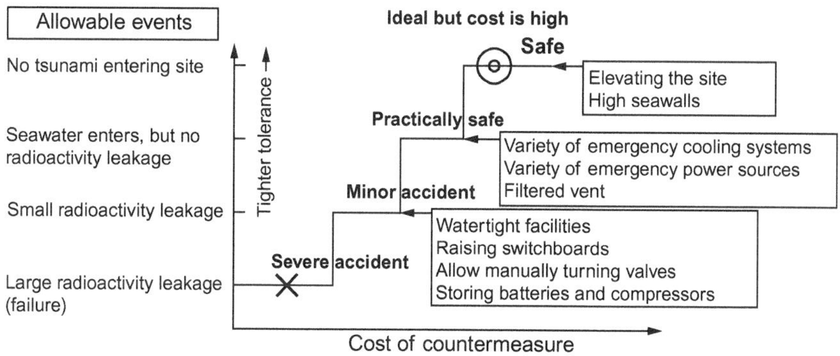

Figure 2.48 Cost of safety measures and allowable severity of accidents. Lower cost countermeasures are available for preventing the worst situation.

2.10.3 Hypothetical scenario for accident prevention

We cannot help wondering, if those on duty had taken different actions under the same circumstances at the time of the accident, such severe consequences would have been avoided. Here we will discuss if the workers and operators had made the best judgments, whether the following scenario of early SRV operation would have been accomplished.

When the reactors were faced with the emergency situation of SBO, while there was enough cooling water in the reactor, if the following steps had been made to allow alternative water injection at low pressure, this severe accident would have been avoided:

(1) Open the SRVs to lower the RPV pressure
(2) Perform vent as necessary to keep the CV pressure low
(3) Continuously inject water using fire engines.

First of all, those on duty must have selected this scenario. Now, opening the SRVs while there was enough cooling water means sudden boiling from the boiling point drop caused by pressure drop. Thus, deciding to follow the scenario requires, first, that the SRVs can be opened, and then confirming that the following vent and alternative water injection with a fire engine can be made for sure. This confirmation requires verifying the availability of fire engines, water source, batteries, compressors, and other tools, feasibility of connecting existing equipment with the emergency tools, checking where the connections will be made, accessibility to those locations, how to execute the vent without power, and securing workforce to accomplish all steps. In reality, all these factors were obstacles in the actual accident response, and they were not, in fact, at satisfactory levels, and those on duty would not have selected this scenario. To answer the question of whether the SRVs would have been opened or not, we will need further investigation, but here we will discuss the authors' views for each reactor.

(1) Unit 1
The IC was not functional after the tsunami attack and at around 20:00 on the first day of the accident, core damage had already started. Thus, even if alternative water injection with a fire engine was the target, *there was hardly any chance for the scenario to be realized.* If, however, the workers had started earlier aiming at the "scenario," alternative water injection from a fire engine and vent could have been made much earlier (several hours) than the records show. If so, the degree of core damage would have been less, and *the amount of radioactivity release could have been much lower. The hydrogen explosion could have been avoided* as well. *Accident response with Units 2 and 3 were greatly affected and delayed with the hydrogen explosion of Unit 1*; thus, if Unit 1 had not exploded, the response for Units 2 and 3 could have been totally different.

(2) Unit 2
Unit 2 had a time slack of over two and a half days after the SBO to the stoppage of RCIC. Thus, this reactor *had a high possibility of executing the scenario* before dawn on March 14th when the core damage started.

(3) Unit 3
Not having batteries and compressors in storage was a big obstacle in the accident response for Unit 3. If, however, there were adequate instructions, securing them could have been completed on March 12th with high possibility of success.

> The accident response for Unit 3 was carried out simultaneously with that for Unit 1; thus, we also have to evaluate restrictions on manpower. The authors, however, do not have enough information to judge such points. We then conclude that for Unit 3, if the right decisions and instructions were made immediately after the SBO, though with some uncertainty, *the chance cannot be denied that the scenario would have been realized.*

From the preceding discussion, when we turn our discussion to Fukushima-1 as a plant, we believe there was a possibility that a severe accident broke out only with Unit 1.

2.11 Background to nuclear power generation

We wrote this section for readers without background knowledge of nuclear power generation so they can better understand this book. The subjects we will cover are as follows:

- Basic knowledge of nuclear fission and decay
- Details of cooling systems and SRV
- Explanation of mechanism of water level indicator malfunction

2.11.1 Nuclear fission and decay

Fragments of material that nuclear fission produces are called fission products, and they are generally unstable immediately after the fission. These products, as Figure 2.49 sketches, produce heat (decay heat) and radiation (α-ray, β-ray, and γ-ray) to slowly change into a stable material that does not produce radiation.

When an operating reactor is suddenly stopped, the reactor holds a large amount of unstable fission products, and the decay heat is as large as 6% of the heat generation during operation. The decay heat gradually drops as the fission products turn into stable materials.

Fuel in a nuclear reactor, thus, continues to produce heat after the operation is stopped (Figure 2.50). The nuclear reactor has RHR systems to cool the fuel even after the reactor is stopped. As time passes, the decay heat drops, but it does not go down to

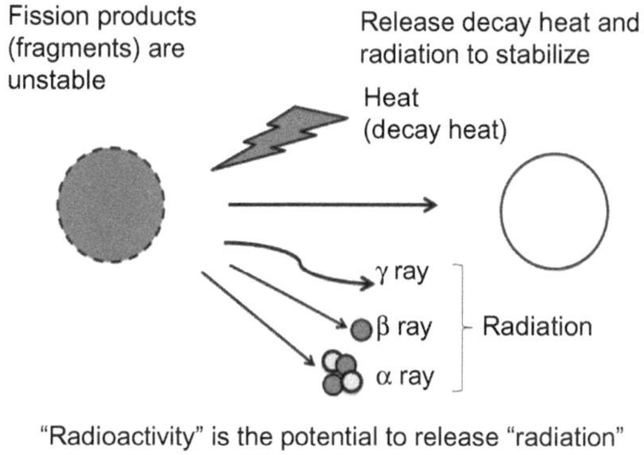

Figure 2.49 Generation of radiation and heat upon decay of fission products.

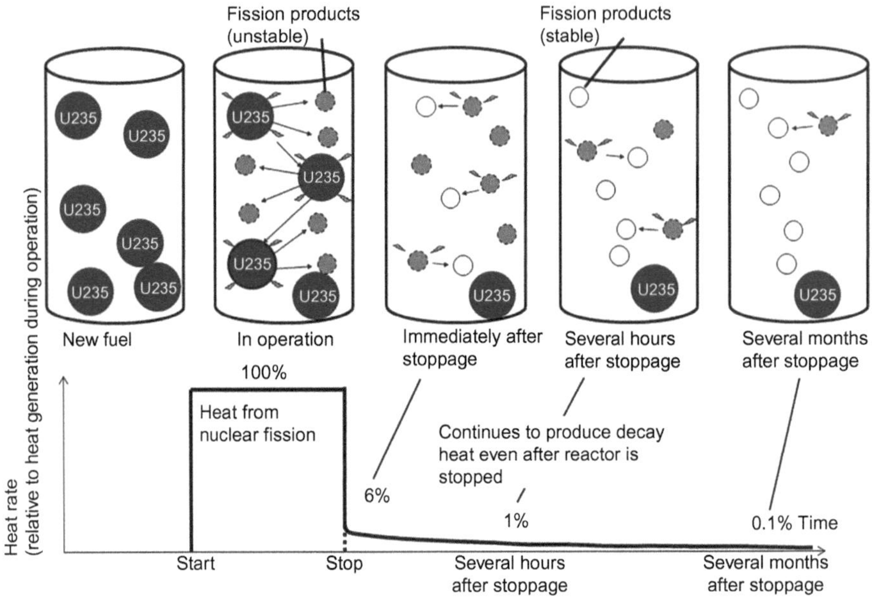

Figure 2.50 Stabilization of fission products and decay heat transition after stoppage.

zero. This is why the fuel, after a certain amount of time left cooled in the reactor, is moved to the fuel pool for storage and natural cooling.

Radioactive materials that nuclear power plants generate are fission products from nuclear fission chain reaction and transuranium elements generated when uranium captures neutrons (Table 2.7). Nuclear fission takes several forms of separation, and a number of fission products make the fragments. Uranium with a mass number of 235 (U235) has a high probability of separating into a material with a mass number of around 130 and another of around 90 (Figure 2.51).

When non-fissile U238 captures a neutron not consumed for fission, it adds to its atom number and mass number to transfer to Plutonium Pu239. Also, when Pu239 captures two neutrons, it transforms to Pu241, and furthermore, when Pu241 decays, it transforms to Amersham Am241. These materials with atom numbers larger than uranium are called transuranium elements, and they are radioactive materials also produced in a nuclear reactor.

Table 2.7 shows main radioactive materials that are produced inside the reactor of a nuclear power plant in operation [7]. Among these materials, volatile radioactive

Table 2.7 **Radioactive material produced in a nuclear power plant**

Category	Source	Primary nuclei
Volatile Quasi-volatile	Nuclear fission	I131, I133, Xw133, Kr85 (noble gas) Cs134, Cs137, Sr90

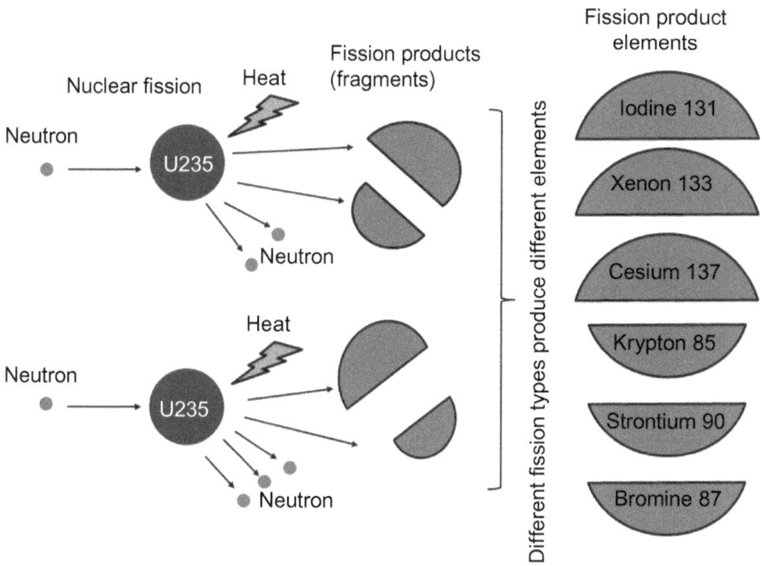

Figure 2.51 Uranium fission and fission products.

materials like iodine I and cesium Cs easily make their ways to the outside when the fuel is damaged.

2.11.2 Cooling systems

Figure 2.52 shows the list of Fukushima-1 NPP cooling systems categorized with their functional requirements.

Table 2.8 shows the primary specifications with these systems.

2.11.3 Safety relief valve (SRV)

The SRV has two functions: to automatically release steam when it reaches a preset value (safety valve function) and to manually release steam when triggered by an external force (relief valve function). Figure 2.53 shows its structure. SRVs discharge steam into the S/C. The reactor also has dedicated safety valves that discharge steam to the D/W. As Table 2.9 shows, the discharge set points are lower for the S/C than the D/W. This is to discharge primarily to the S/C so radioactive material is captured by water in the S/C pool.

A nuclear reactor may require lowering its pressure, to a value lower than the "safety valve" function settings to allow additional water injection from the outside. For such situations, SRVs also have a "relief valve" function that is force activated by control signals from the central control room. When opening SRVs, if the CV pressure (back pressure to the piston) is higher than the design pressure, larger forces may be required to activate the relief function even if the RPV pressure is the same. The

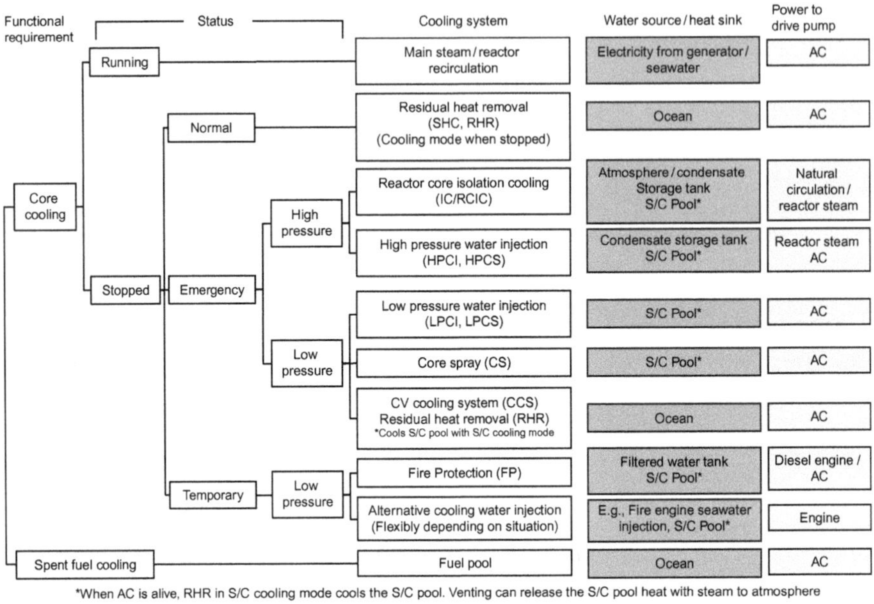

Figure 2.52 Fukushima-1 NPP cooling systems and their categories.

Table 2.8 **Primary specifications of fukushima-1 NPP cooling systems (*adapted from [1], [11]*)**

Fukushima-1 NPP		Unit 1	Unit 2	Unit 3
Shutdown cooling system (SHC)	Pump			
	No.	2		
	Flow rate (m^3/h/pump)	465.5		
	Head (m)	45.7		
	Heat exchanger (EX)			
	No. of units	2		
	Capacity (kcal/h)	3.80×10^6		
RHR	Pump			
	No.		4	4
	Flow rate (ton/h)		1750	1820
	Total head (m)		128	128
	Seawater pump			
	No.		4	4
	Flow rate (m^3/h)		978	978
	Total head (m)		232	232

Continued

Table 2.8 **Continued**

Fukushima-1 NPP		Unit 1	Unit 2	Unit 3
	Heat EX			
	No.		2	2
	Heat Xfer rate (kcal/h)		7.76×10^6	7.76×10^6
IC	No. of systems	2		
	Tank capacity (m^3/tank)	106		
	Steam Fl Rt (ton/h/tank)	100.6		
RCIC	Steam turbine			
	No.		1	1
	RPV pressure (Mpa)		7.7–1.0	7.7–1.0
	Output (HP)		500–580	500–580
	Rotation (rpm)		5000–2000	4500–2000
	Pump			
	No.		1	1
	Flow rate (ton/h)		95	97
	Total head (m)		850–160	850–160
	Rotation (rpm)		Variable	Variable
HPCI	No. of systems	1	1	1
	Flow rate (ton/h)	682	965	965
	No. of pumps (per system)	1	1	1
LPCI	No. of systems		2	2
	Flow rate (t/h)		1750	1820
	No. of pumps (per system)		2	2
Core spray system (CS)	No. of systems	2	2	2
	Flow rate (ton/h)	550	1020	1141
	No. of pumps (per system)	2	1	1
	Pump discharge pressure (MPa)	2	3.45	3.45
CV cooling system(CCS)	No. of systems	2	2	2
	Dsgn Fl Rt (ton/h/system)	705	2960	2960
	No. of pumps (per system)	2	2	2
	No. of heat Ex (per system)	1	1	1

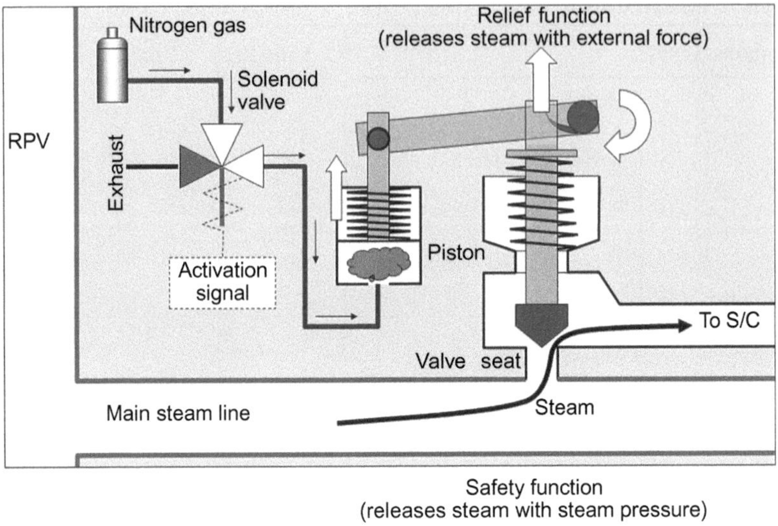

Figure 2.53 Sketch of SRV (relief valve function and safety valve function).

driving force to activate the relief function is nitrogen gas at pressure of less than 1 MPa. During the Fukushima-1 accident, the CV pressure often exceeded 0.5 MPa. This backpressure may have made the relief functions hard to activate.

Table 2.9 **Fukushima-1 SRV specification** ***(adapted from [1], [11])***

Fukushima-1 NPP		**Unit 1**	**Unit 2**	**Unit 3**
Main steam line SRV	Number of valves	4	8	8
	Total capacity [ton/h]	1090	2900	2900
	Relief discharge pressure [MPa] (No. of valves)	7.28(1) 7.35(2) 7.41(1)	7.44(1) 7.51(3) 7.58(4)	7.44(1) 7.51(3) 7.58(4)
	Safety discharge pressure [MPa] (No. of valves)	7.65(2) 7.71(2)	7.65(2) 7.71(3) 7.79(3)	
	Discharge to	S/C	S/C	S/C
Dedicated safety valve	Number of valves	3	3	3
	Total capacity [ton/h]	900	900	900
	Discharge pressure [MPa]	8.51(2) 8.62(1)	8.55	8.55
	Discharge to	D/W	D/W	D/W

2.11.4 Reactor water level sensor and mechanism of malfunction

Principle of nuclear reactor water level sensing

The principle of nuclear reactor water level sensing is to measure the water pressure change at the bottom of the pressure vessel to detect the cooling water level (Figure 2.54). If there was a water level change of 2 m, the change in pressure then is about 0.02 MPa.

If the water level sensor simply measures water pressure at the bottom of the pressure vessel as shown in Figure 2.54a, the measured pressure is the cooling water pressure (assume the water level at about 10 m) 0.1 MPa plus the steam pressure of about 7 MPa to bring the reading to about 7.1 MPa. Then the measurement range is too large for accurately measuring changes around the size of 0.02 MPa.

The sketch in Figure 2.54b solves this problem by measuring the difference between the steam pressure at the top of the pressure vessel and the water pressure at the bottom of the reactor. This difference shows the reactor water level at accuracy several tens of times better.

The water level sensors in actual nuclear reactors are as Figure 2.54c shows. The reference condensate pot at the top of the reference side of the pressure gauge allows measuring the difference between the reactor water level and the water level in the reference condensate pot. This configuration further improves the accuracy of water level measurement.

This configuration, however, as we will discuss later, has a fundamental problem that if the water level in the reference condensate pot goes down with the RPV temperature rise, the water level reading goes wrong. Some countermeasures against this problem should be made for the case of severe accidents.

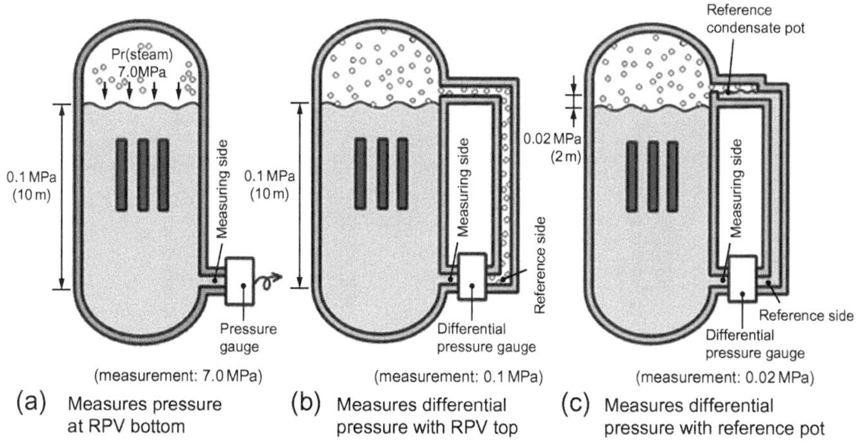

Figure 2.54 Principle of nuclear reactor water level measurement.

Mechanism of malfunction

Normal state

Assume the reference pot water level was 10 m, reactor water level was 8 m, and the steam pressure was 7 MPa (Figure 2.55). The steam pressure of 7 MPa and the water pressure 0.1 MPa from the reference pot to the pressure gauge apply to the reference side of the pressure gauge at 7.1 MPa. On the measurement side, the sum of the steam pressure 7.0 MPa and cooling water pressure 0.08 MPa apply at 7.08 MPa. The sensor then measures the pressure difference of $\Delta h = 0.02$ MPa, which converts to a water level of 2 m.

Reference pot water level drop

When the CV temperature rises above the saturate temperature or the temperature around the reference condensate pot goes up and condensation no longer takes place, the water in the reference side piping evaporates and the water level in it drops. The sensor, unaware of the change, bases its reading on the state with water inside the reference condensate pot, and the reactor water level is miscalculated to be higher than actual (Figure 2.55b). The sensor without a reference condensate pot shown in Figure 2.54b does not have the problem of evaporation, but if steam condenses inside the piping, the water in the reference piping causes the same problem of incorrect reading.

Extreme water level loss in the reactor

If the reactor water level drops drastically to below the inlet of the measuring side piping, the pressure sensor is merely measuring the elevation difference of water inside the piping (Figure 2.55c). This reading is not relevant to the reactor water level.

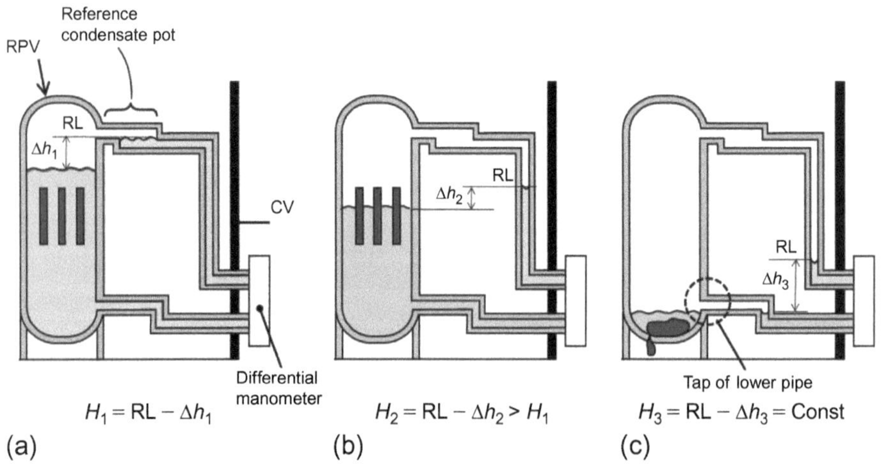

Figure 2.55 Malfunction of RPV water level gauge. (a) Normal state, (b) evaporation lower water level in reference side, and (c) significant water level drop in the reactor.

If the system was in that state and if evaporation in the piping on the two sides stops, the indication stabilizes at a wrong value.

Data show that during this accident, such phenomena took place in Units 1 through 3. Those in charge, however, did not recognize that the reactor water level readings were incorrect, and judgments made without knowing the actual state of the reactors were also root causes of this accident.

2.12 Conclusions

Unit 1 of Fukushima-1 was one of the first, and thus the oldest, commercial BWRs in operation. Units 2 and 3 designs were newer with bigger capacity and thus with newer emergency systems at the time of core isolation (RCIC). Although RCIC for Units 2 and 3 kicked in after the earthquake and kept running after the tsunami attack, the older IC for Unit 1 that started after the earthquake was forced to stop by its own logic on loss of electricity caused by the tsunami.

The operators of the plant did not recognize that the IC had stopped. It was the fail-safe function, which nobody remembered at the time of the accident, to close the valves in case of power loss that shut down the IC. This oversight can be excused due to the chaotic situation then; however, not ever test running the IC system since the plant start-up some 40 years earlier contributed to the problems. None of the operators or engineers had experience of witnessing the IC system in operation.

The lack of preparation against an extended SBO and simultaneous accidents with multiple reactors was evident. Without ways of reading important parameters like reactor water level and no way of mechanically forcing valves that were motor- or air-driven under normal circumstances, plant personnel had to work with batteries, portable compressors, and generators. This emergency equipment reached their hands when it was too late to bring the accident under control. It was, however, the dedicated hard work of these men that saved the accidents from turning into a worse disaster that could have threatened the lives of people in the Tokyo metropolitan area.

The images of explosions of Units 1 and 3 shook the entire world that was keeping an eye on the news as the accident progressed. It was, however, Unit 2 that released the biggest amount of radioactivity. The accident was a severe reminder about how vulnerable modern machines are against loss of electricity. The Japanese design approach which focused on preventing even the smallest amount of radiation release led to the inability to prevent a more catastrophic event.

References

[1] Investigation Committee on the accident at the Fukushima Nuclear Power Station of Tokyo Electric Power Company, Interim report. http://www.cas.go.jp/jp/seisaku/icanps/eng/interim-report.html, 2011 (accessed 08.08.14).

[2] Investigation Committee on the accident at the Fukushima Nuclear Power Station of Tokyo Electric Power Company, Final report. http://www.cas.go.jp/jp/seisaku/icanps/eng/final-report.html, 2011 (accessed 08.08.14).
[3] NHK Broadcast (in Japanese), Fukushima Nuclear Power Plant accident as viewed from the world. Japanese reference page URL http://www.nhk.or.jp/documentary/1203.html, 2012 (accessed 08.08.14).
[4] NHK Broadcast (in Japanese), Fukushima Nuclear Power Plant accident as viewed from the US. Japanese reference page URL https://www.nhk.or.jp/etv21c/file/2011/0814.html, 2011 (accessed 08.08.14).
[5] NHK Broadcast (in Japanese), Meltdown—what happened in Fukushima-1 then. Japanese reference page URL http://www.nhk.or.jp/special/detail/2011/1218/, 2011 (accessed 08.08.14).
[6] Diablo Canyon Power Plant in light of the incident at Fukushima Daiichi Nuclear Power Plant, 2011 presentations, Session 27, PGandE, National Radiological Emergency Preparedness Conference. http://www.nationalrep.org/presenters_main/Inventory2011.aspx, 2011 (accessed 08.08.14).
[7] Nuclear Power Handbook (in Japanese), Editing committee for Nuclear Power Handbook, Ohm, 2007.
[8] Interim report: IV-3.-(2)-a (p. 122, line 20–22 in the Japanese, p. 143, line 4–6 in the English version). http://www.cas.go.jp/jp/seisaku/icanps/eng/120224Honbun04Eng.pdf.
[9] Interim report: IV-3.-(1)-b.-(iii) (pp. 106L8–107 in the Japanese, pp. 125L9–126 in the English version). http://www.cas.go.jp/jp/seisaku/icanps/eng/120224Honbun04Eng.pdf.
[10] Interim report: IV-3.-(1)-e.-(b) (pp. 111–114 in the Japanese, pp. 132–134 in the English version). http://www.cas.go.jp/jp/seisaku/icanps/eng/120224Honbun04Eng.pdf.
[11] Fukushima Nuclear Accident Analysis Report, June 20, 2012, TEPCO, http://www.tepco.co.jp/en/press/corp-com/release/betu12_e/images/120620e0104.pdf, (accessed 08.08.14).

The response of central and local government agencies to the Fukushima nuclear power plant accident

3

S. Abe
Kansai University, Osaka, Japan

3.1 Introduction

The Fukushima-1 Nuclear Power Plant (NPP) accident was triggered by the combined natural disasters of earthquake and tsunami; however, the Tokyo Electric Power Company's (TEPCO) insufficient preparation against accidents and disasters, its weakness in managing the accident, and deficiencies in government nuclear disaster preparedness turned the accident into a serious nuclear disaster of International Nuclear Events Scale (INES) Level 7 (see comments on TEPCO's performance in Refs. [10–12] as well as comments on the government's response in Refs. [13–16]; see also Refs. [17,18]).

In terms of measures for accident and disaster prevention, the Nuclear and Industrial Safety Agency (NISA) and TEPCO's preparations against tsunami and severe accidents were insufficient, lacked precautions against large-scale complex disasters, and did not have disaster prevention or evacuation plans in case of Containment Vessel (CV) damage and a large radiation release from the plant.

TEPCO, after the accident, made a number of serious errors in the field. Both the central and the local government are also responsible for a share of blame due to the lack of plans for preventing magnification of damage, monitoring, making use of the System for Prediction of Environmental Emergency Dose Information (SPEEDI), giving effective evacuation directions to the residents, taking measures against exposure to radiation, and making public announcements to the people of Japan and overseas (see Refs. [19–21] for comments on government response and also see Refs. [17,18]).

This chapter discusses the problems of the central and local government agencies. The appendix at the end of this chapter lists all the organizations and people that are mentioned in this book. The indentation indicates the organizational hierarchy and the person's affiliation.

The 2011 Fukushima Nuclear Power Plant Accident. http://dx.doi.org/10.1016/B978-0-08-100118-9.00003-6

3.2 Central government: key responsibilities and preparatory measures

3.2.1 The "defense in depth" strategy

The International Atomic Energy Agency (IAEA) was created in 1957 to promote the use of nuclear energy for peaceful purposes and to prevent its misuse for weapons. IAEA has been promoting the concept of "defense in depth" to ensure safety within nuclear power facilities.

In 2006, IAEA, joined with eight other international organizations, including the European Atomic Energy Community (EURATOM), the OECD Nuclear Energy Agency (OECD/NEA), and the World Health Organization (WHO), to consolidate three documents about safety principles, "The Safety of Nuclear Installations," "The Principles of Radioactive Waste Management," and "Radiation Protection and the Safety of Radiation Sources," and establish "Fundamental Safety Principles" into a coherent and consistent set of ten new principles [1]. Principle 8 of this set of ten new principles states the primary means of preventing and mitigating the consequences of accidents is "defense in depth."

IAEA states, "The primary means of preventing accidents in a NPP and mitigating the consequences of accidents if they do occur is the application of the concept of defense in depth. This concept is applied to all safety-related activities, whether organizational, behavioral or design related, and whether in full power, low power or various shutdown states" [1]. This "defense in depth" has the following five levels.

Defense Level 1 Defense at this level aims at preventing deviation from normal operation and also preventing safety-related malfunctions and failures. For this purpose, commissioning, design, construction, maintenance, and operation have to proceed in a sound and conservative manner.

Defense Level 2 The purpose of defense at this level is to detect anticipated operational occurrences deviating from normal operation and to prevent a situation from developing into an accident. For this purpose, the design has to prepare special systems and mechanisms, and their effects have to be verified with safety analyses to prevent such occurrences or minimize their effects, and furthermore, operational procedures have to be in place to return the plant to a safe state.

Defense Level 3 A postulated initiating event during operation may develop into a situation uncontrollable with Defense Level 2 and into an accident. The plant design shall take such accidents into account and equip the plant with safety mechanisms, safety systems, and procedures to return the plant to a safe state and prevent core damage and radioactivity release to the surrounding area outside the plant.

Defense Level 4 Defense at this level ensures the confinement function by reliably keeping the radiation release to the lowest possible level to mitigate various accident effects from failure in Defense Level 3.

Defense Level 5 Defense at this level targets mitigating the radiation effects from radioactivity release caused by an accident state. Defense calls for preparing a well-equipped emergency control center, as well as emergency plans and emergency procedures within the area and also in surrounding regions to counter emergency situations.

Table 3.1 Summary of defense in depth

Defense level	Target/purpose of defense
1	Prevent deviation from normal operation
2	Detect anticipated operational occurrences and prevent their development into accidents
3	Moderate effects of accidents from postulated initiating events
4	Action against severe accidents
5	Countermeasures against radiation release caused by accidents

Table 3.1 summarizes these five levels of defense. In Japan, such defense in depth is also called multilayered protection; however, in Japan the fourth and fifth levels had never gone through serious evaluation. Both the regulators and the industry only took measures up to Defense Level 3 for their defense in depth.

In other words, guidelines for safety design reviews and technical standards have addressed, to some extent, nuclear plant defense up to Level 3; however, as we will later describe, the higher Levels of 4 and 5 have only been partially planned: Defense Level 4 through evaluation and application of accident management as instructed by regulation, Defense Level 5 within the framework of "Act on Special Measures concerning Nuclear Emergency Preparedness" [2]. The preparations in place were not at the level IAEA required.

Fukushima-1 had some preparations in place for Levels 1 through 3. For example, by 1999, the plant had added diesel generators and evaluated installing air-cooled generators. The plant made these improvements as part of accident management in response to Level 3 as we will discuss later. These steps, however, were not a plant design of measures against situations where uncontrollable initial events could turn into accidents; thus, the modifications cannot claim to be defense in depth preparations to Level 3.

3.2.2 Assessing the performance of the main government departments in safety regulation and disaster management

This section reviews the problems with the Japanese regulatory bodies in terms of Defense Level 4.

Nuclear reactor facilities by design have multiple levels of countermeasures against unexpected incidents and accidents. The 1979 Three Mile Island (TMI) and 1986 Chernobyl accidents, however, have shown us that severe accidents that damage the reactor core beyond the design standards can happen.

Thus, in the 1980s and 1990s, international discussions began about how to counter such severe accidents. The discussions led to international acceptance of the basics of ensuring NPP safety so that licensing is granted to safety assurance to within the framework of design standards of nuclear reactor facilities, and serious accidents that exceed the design standards and damage the core or nuclear fuel are handled through measures against severe accidents. Each country then established its own severe accident measures.

At this time, Japan started discussions about severe accident measures as well; the Nuclear Safety Commission, which had been analyzing the TMI and Chernobyl accidents, set in 1987 a Common Problem Round-Table Group within the Special Committee on Nuclear Reactor Safety Standards to evaluate measures against severe accidents. The group issued a report in March 1992. Based on this report, NSC announced, two months later, its decision "Accident Management: Measures against Severe Accidents at Light Water Nuclear Reactor Facilities for Power Generation" [3].

The decision stated "Events that exceed the design standards may rarely cause large damage to the core, and Accident Management are measures that prevent the development of such events into severe accidents or even if they have developed into severe accidents, moderate their effects with safety margins within the current design, with accountable functions that are beyond their original functions, or with effective use of newly installed equipment in preparation to such situations" [3].

The preceding definition in Japanese did not contain the Japanese word for "severe accident" and instead used the phonetic sound of the English phrase "severe accident" in the Japanese alphabet. Calling this definition "Accident Management" suggested an assumption that severe accidents were unlikely.

This decision by the Nuclear Safety Commission set the later fundamental direction in Japan in terms of severe accident measures and accident management. These are summarized as follows:

(1) Safety of Japanese nuclear reactor facilities has complete assurance based on the concept of multilayered protection and strict safety assurance measures. The possibility of severe accidents is negligibly small from an engineering standpoint, and risks with nuclear reactor facilities are sufficiently small.
(2) Preparing accident management measures further lowers this already low risk; thus, each nuclear reactor owner is encouraged to make voluntary efforts to prepare them.

In other words, the decision implied that the possibility of severe accidents in Japan is very small, and voluntary efforts by the utility companies would suffice in dealing with them. Chapter 4 will discuss problems with TEPCO's accident management.

This decision was questioned after the Fukushima NPP accident, and the Nuclear Safety Commission, on October 20, 2011, abandoned it. In the statement then issued, the Commission Chair stated as follows:

> *The occurrence of this accident revealed that we had serious problems with our idea that "risk is sufficiently low" and with the effectiveness of voluntary efforts by the utility companies in further lowering the risk. The definite problem lied in the fact that despite the large risks of external events had been pointed out, especially with earthquakes and tsunami, actual measures failed to fully account for such risks. Preparing accident management for all nuclear reactor facilities took a total of 10 years,*

and they had their bases in countermeasures against occurrences identified by probabilistic risk analysis of internal events made in 1994 and the bases had never been reevaluated. Furthermore, no one recognized the possible failure of equipment or procedures for accident management.

3.2.3 Assessing the use of severe accident measures

When the Nuclear Safety Commission was discussing severe accident measures, the Ministry of International Trade and Industry (MITI, now merged into METI) regulated the safety of nuclear reactors for power generation. After the decision by the Nuclear Safety Commission, the ministry composed "Future Processes of Accident Management" [4]. At the same time, it announced a notice to business directors of public utilities titled "Preparing Accident Management in Nuclear Power Plants" [5]. This notice acknowledged the need for severe accident measures; however, it was in line with the commission's understanding that accident management as a measure against severe accidents was up to voluntary actions by each utility company.

The following circumstances had MITI take the stance that the utility companies were to carry out independent actions for accident management.

Fundamentally, those involved had the mind-set that regulations at the time did in fact provide sufficient safety for Japanese nuclear reactors. This appeared to be supported by probabilistic safety assessment (PSA). PSA is a method for overall quantitative evaluation of safety of nuclear reactor facilities by quantitatively analyzing frequencies of events that lead to abnormality or accidents within nuclear reactor facilities, probability of loss of safety functions for mitigating the effects of such events, and the processes and effects of these events. The method concluded that the probability of severe accidents occurring in Japan were 10^{-6}/reactor-years, a probability less than the IAEA target at the time of 10^{-4}/reactor-years for old reactors and 10^{-5}/reactor-years for new ones. The administration thus believed that the regulation at the time was sufficient to ensure safety.

Another potential consideration was to avoid the risk of complicated lawsuits. Ever since NPP construction became more frequent in a number of regions in Japan, a number of lawsuits had been filed against the government seeking cancellation of the licensing of the nuclear reactors. The government claimed the regulation back then was sufficient for safety assurance of nuclear reactors. The administration judged that if they required new laws and regulations to take actions for severe accident management, it might be seen to admit flaws in regulations at the time, and facilities that had been built previously would face difficulties with ongoing litigation.

There are, however, two types of factors that can lead to severe accidents that exceeded existing design standards: internal factors and external factors. Internal factors are troubles within the NPP (e.g., failure of equipment or human error by the operator), and external factors include possible natural phenomena like earthquakes, flooding, tsunami, wind, freezing weather, snow, and landslides, in addition to external human-caused factors like airplane strikes, destruction of dams, or explosion [17,18,22].

Severe accident measures counter these events that exceed the design standards, and thus, the internal and external factors should have been evaluated separately.

When MITI started evaluation of accident management, it had the intention to first start with internal factors of machine failures and human error and then to continue with external factors like earthquakes. As MITI discussed the evaluation with utility companies, external factors were pushed back for later review. Accident management by utility companies, thus, only addressed internal failures and human errors in the area of severe accident measures (Figure 3.1).

This is how the regulatory organizations and utility companies developed the belief that NPPs have sufficient safety measures and did not pursue the severe accident measures of accident management on external factors. As a result, Fukushima-1 lacked proper preparation against a tsunami-induced complete loss of electrical power and failed to cool its nuclear reactors.

The accident has shown us that severe accident measures should not have been left to voluntary actions by utility companies but instead, should have been managed with appropriate laws and regulations.

3.2.4 Assessing the response to extended station blackout (SBO)

Countermeasures against SBO are among the most important of severe accident measures. Fukushima-1 lost all its external power supplies and most of the internal power supplies, lost the core cooling functions, and headed into a severe accident.

SBO is a state in which all external electrical power supplies and on-site electricity supplies from all emergency sources are lost. It is an extremely serious threat to maintaining the safety of NPPs. Thus, in Japan, the Review Guide for Safety Design by the Nuclear Safety Commission (formerly the Atomic Energy Commission up to 1978) had listed securing electrical power sources as a requirement in the Guidelines. The Atomic Energy Commission document, "Review Guide for Safety Design of Light Water Nuclear Power Reactor Facilities" [25], made public in June 1977, stated Guideline 9 as follows:

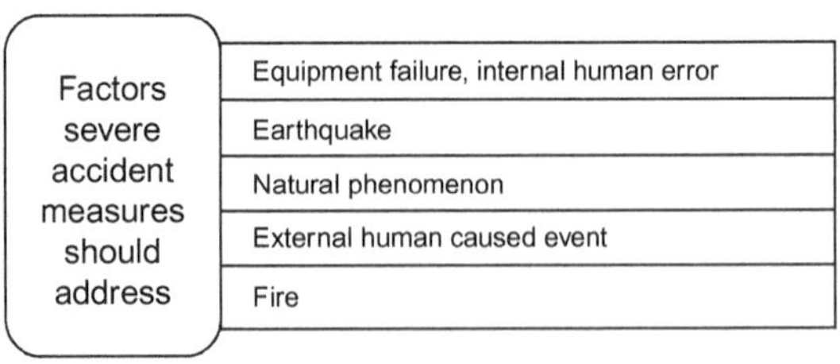

Figure 3.1 Areas of severe accident measures.

Guideline 9: design checkpoint against power outage

A NPP design shall, upon a short SBO, safely stop the nuclear reactor and then assure cooling after the stop.

There is, however, no need to plan against a simultaneous functional loss of electric power facilities with their high level of reliability [25].

These guidelines are supported with comments to clarify meanings and interpretation. The comment for Guideline 9 stated as follows:

> *Evaluation of extended SBO is not required because quick recovery of the power line or diesel generators is expected.*
>
> *"High level of reliability" means keeping the emergency power supply system up at all times to avoid startup failure of specific equipment or having multiple units with high reliability available on-site, each with an independent power supply [25].*

The Nuclear Safety Commission, in August of 1990, revised the "Review Guide for Safety Design of Light Water Nuclear Power Reactor Facilities," and the guideline for cases of power outage was moved to Guideline 27, which stated as follows:

Guideline 27: design checkpoint against power outage

Nuclear reactor facilities shall be designed against a short SBO to stop the nuclear reactor safely and ensure cooling after such stopping [23].

The comments explained Guideline 27 as follows:

> *Evaluation of an extended SBO is not required because quick recovery of the power line or diesel generators is expected.*
>
> *When the emergency AC power supply facility has high reliability with its system structure or operation (e.g., by keeping it up and running at all time), SBO does not need to be evaluated in design [23].*

The safety design review guideline by the Nuclear Safety Commission is used for double-checking, and its articles, strictly speaking, are not laws or regulatory requirements for utility companies. If a company, however, is not following them, it will have a hard time passing the review by this regulatory organization. For utility companies, in reality, these articles were equivalent to laws or regulatory requirements. This specific Guideline 27, however, had the following problem.

The Atomic Energy Commission and the Nuclear Safety Commission interpreted "short" in Guideline 27 to mean 30 min or less since 1977. This understanding caused utility companies to interpret the capacity requirement on chargeable batteries to last at least 30 min in case of SBO.

The guideline was written under the assumption that failure of off-site power sources and on-site supplies were independent, and simultaneous failure of both supplies and damage on the switchboard had never even been evaluated. Such assumptions were clearly an oversight on the part of those involved with editing this safety design review guideline.

In the earlier safety reviews, some technical experts had repeatedly asked about the reason for the setting of "short time" to 30 min and why there was no evaluation of extended SBO during nuclear safety committees; however, the discussions never led to doubts about the reviews themselves or adequacy of the guidelines. The guideline saying that "an extended SBO does not require evaluation" remained for years.

3.2.5 Assessing the countermeasures in the event of a tsunami

Fukushima-1 NPP was, at first, licensed with a design wave height of 3.1 m. Emergency seawater pump facilities for Units 1 through 4 were built on a 4-m bedrock and the reactor and turbine buildings on 10 m. The wave height of 3.1 m was based on the tsunami from the 1960 Chile earthquake. The construction was doomed to lose seawater cooling with a tsunami runup height of 4 m and DC power and emergency diesel generators with a runup height of 10 m.

TEPCO later reviewed the expected tsunami height and revised the highest wave attack to Fukushima-1 at 5.7 m and later at 6.1 m. In 2002, the emergency seawater pumps were lifted as we later describe in Chapter 4. This later work ensured that the emergency seawater pumps would maintain their cooling function to prevent core damage even with the loss of other facilities on the 4 m bedrock. In reality, however, the tsunami from the Tohoku Area Pacific Offshore Earthquake reached a height above 10 m, and the reactor cooling function was lost.

In July 1993, the Southwest Hokkaido offshore earthquake occurred, and a number of tsunami waves affected Okushiri Island and other areas. The severe damage from this tsunami led MITI to order reevaluation of safety against tsunami to existing nuclear power facilities. TEPCO replied by submitting "Fukushima 1 and 2 Tsunami Safety Evaluation Report" to MITI. This indicates that regulatory bodies had recognized the risk of tsunami to some extent since the 1990s.

The Nuclear Safety Commission, with responsibility for setting the guidelines for the safety design review of NPPs, had plans to include articles about countermeasures against tsunami as phenomena that accompany earthquakes in their guidelines. The commission's committee administration office, also had recognized the need to discuss tsunami in the guidelines.

Before taking on the revision work, thorough discussions about earthquake preparation took place in the Nuclear Power Engineering Corporation (NUPEC, closed in March of 2008 with its work transferred to JNES and The Institute of Applied Energy); however, no independent discussions on earthquakes were held, and no tsunami specialist was in the antiseismic work group. A tsunami wave was understood as a phenomenon that may follow an earthquake, and once its origin is set, a computer simulation can calculate the wave height; thus, they saw no need for a tsunami expert.

The history of tsunami, however, with information on damage caused, countermeasures taken, and their characteristics, are hard topics for seismic specialists. The lack of a tsunami expert in the work group clearly reveals the limited understanding of the Nuclear Safety Commission members at the time about the significance of possible problems with a tsunami.

The revision of the Guidelines for Anti-Seismic Design Review by the Nuclear Safety Commission took five long years to complete in September of 2006 [6]. The last chapter of the final version was "8. Discussions on Phenomena that Accompany Earthquakes," including landslides at slopes surrounding facilities and tsunami. The requirement for tsunami preparation stated, "Although extremely rare, a tsunami is possible while the facilities are in service, and thus assuming its occurrence is adequate. Facilities shall be designed so their safety functions are not severely affected even upon a tsunami attack." It was a step forward that the guidelines mentioned countermeasures against tsunami; however, the sentence did not trigger any new tsunami preparation, and TEPCO and other utility companies did not make any major changes to their tsunami preparations.

3.3 The response and preparedness of the central government

3.3.1 *Assessment of emergency measures from the Nuclear Emergency Preparedness Act guidelines*

Lessons from the 1999 JCO criticality accident in Tokai, Ibaragi prefecture, led to the issuance of the "Act on Special Measures concerning Nuclear Emergency Preparedness (Nuclear Emergency Preparedness Act)" in the same year to enhance the response to nuclear power accidents. The law called for a local emergency response headquarters with a commander armed with authority delegated by the prime minister. The "Nuclear Disaster Countermeasures Manual (Nuclear Emergency Preparedness Act Manual)" was also made available. The manual was based on this law and was constructed under the condition that the local headquarters would play a central role.

We next review whether these systems for countering nuclear disasters functioned as designed during the Fukushima-1 NPP accident.

At 19:03 on March 11, 2011, the government declared a Nuclear Emergency Situation and set up the Nuclear Emergency Response Headquarters (NERHQ) headed by the Prime Minister Kan in his office, and the local emergency response headquarters in the off-site center in the town of Ohkuma in Fukushima prefecture. The administration office for NERHQ was set in the METI emergency response center. Following the Nuclear Emergency Preparedness Act Manual, METI Vice-Minister Ikeda, TEPCO Vice-President Mutoh, and other primary members gathered at the off-site center to start the local response headquarters.

The off-site center, however, had suffered damage from the earthquake, most of its communication equipment was unusable, and the building had no protection against radioactivity. Because of the difficulty in using the center, the local response headquarters were transferred to the prefectural government office on March 15th. The local response headquarters, thus, failed to function in the initial stage as the emergency command center.

The headquarters in Tokyo had to cover responsibilities originally intended for the local headquarters and be at the very front of the disaster response. Top officials of the ministries and agencies gathered at the crisis management center in the basement of the Prime Minister's office, which had already been set up after the Tohoku Area Pacific Offshore Earthquake. This center in the basement, however, proved inadequate and instead, Prime Minister Kan and the people around him in the fifth floor of the Prime Minister's Office Building (and partially in the concourse level between the first floor and basement) made the critical decisions. Prime Minister Kan himself was at the very front of the response by actively gathering information himself and even flying to the accident site to observe the situation.

The main reason why the Tokyo response headquarters was not used reflected weaknesses in NISA, including receiving proper information from local offices and TEPCO as well as a broader lack of readiness for a disaster on this scale.

To resolve this situation, the Fukushima NPP Accident Response General Headquarters (General Headquarters) was set up inside TEPCO's headquarters main office in Tokyo. Prime Minister Kan was in charge of the headquarters with Minister Kaieda of METI and President Shimizu of TEPCO serving as Assistant Chiefs.

This General Headquarters contributed hugely to the accident response with its teleconferencing system by establishing real-time communication with the Fukushima-1 Plant Response Office to share status and work progress information among the government, TEPCO, and NISA. This General Headquarters in practice served as the emergency response center in place of the government NERHQ.

As explained earlier, the fifth floor of the Prime Minister's Office Building, not specified in the Nuclear Emergency Preparedness Act or Nuclear Emergency Preparedness Act Manual, took the role of the command center and then from March 15th onwards, the role was passed to the General Headquarters with Prime Minister Kan at the forefront (Figure 3.2). The lack of preparation of local response headquarters and NERHQ administration, an insufficient system for collecting information, and poor support from the Nuclear Safety Commission led to such an unplanned situation.

The Nuclear Emergency Preparedness Act, at a glance, seems to be a well-prepared law; however, it was written for a locally containable accident like the JCO criticality accident and was not intended for a complex, large-scale, widely spread disaster as developed this time. Therefore, it was only natural that Nuclear Emergency Preparedness Act and its Manual failed to work.

3.3.2 Response of the regulatory organizations to system failures at the Fukushima NPP

NISA was an organization in charge of safety regulation of power-generating nuclear power facilities and, at the same time, was responsible for taking the central role in disaster response in case of nuclear power accidents with commercial reactors by administering NERHQ.

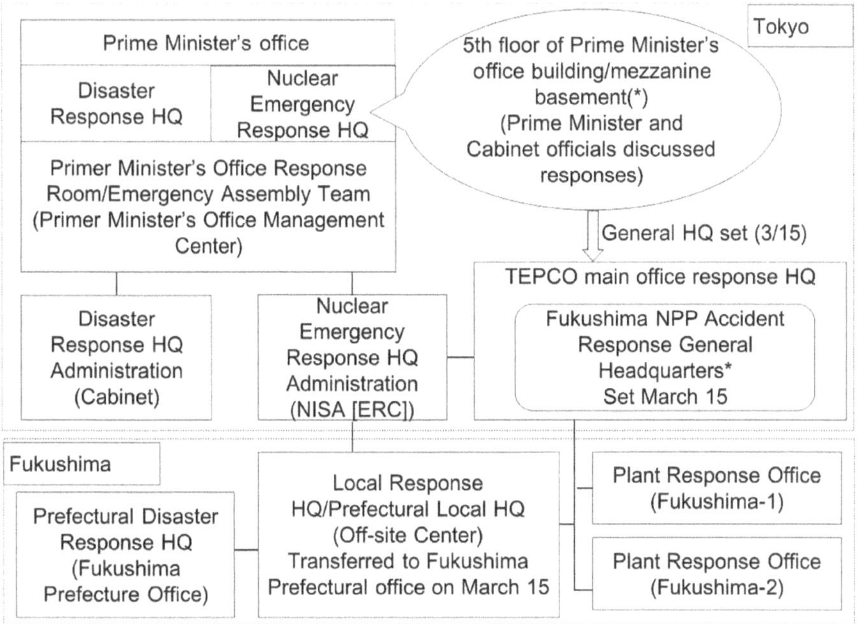

Figure 3.2 Actual Response Organization for Fukushima-1 NPP Accident.
Source: Final Report, p. 198.

NISA, however, at time of the Fukushima-1 NPP accident, displayed weakness in a number of aspects:

(i) It failed to effectively collect information to transfer to the Prime Minister's Office and related ministers and agencies.
(ii) It did not explain sufficiently the status, progress, necessary measures, and articles of Nuclear Emergency Preparedness Act to the Prime Minister's Office in case of nuclear power disasters.
(iii) Although having radiation spread information from SPEEDI available, it interpreted it as inapplicable without the radiation source information and did not make use of it.
(iv) It failed to perform its duties as the administration office, such as delegating authority to the local response headquarters.

NISA also did not have input in working out mid- to long-term projects for preventing nuclear disasters by safety regulations for severe accident responses. The result was that it failed to order the utility companies to take adequate severe accident measures.

Knowledge and problems relating to nuclear safety are closely related to the field; thus, keeping up with the skills for safety assurance may be difficult for the regulating body. Practical and specialist knowledge about safety and technology at the level of a utility company and ability in making difficult decisions and performing tasks are required. Such qualifications should be posed not only on individuals in the workforce; organizational and systematic functions for safety assurance are also needed. NISA, in addition to the aforementioned shortcomings, lacked sufficient resources in this respect.

The lack of protection against radiation at the off-site center clearly shows NISA's lack of preparation. An off-site center is where the local response headquarters will be set to take the central role in emergency situations in case of nuclear disasters. The off-site center for Fukushima-1 was located in the town of Ohkuma about 5 km from the accident site.

The center, located so close to the NPP, did not have an air-cleaning filter to block radioactive material in the air. On March 14th, after the reactor building of Unit 3 blew up lifting the level of radiation, staff had to abandon the off-site center. In other words, this was a facility prepared in case of a nuclear disaster; however, its structure lacked any preparation against a rise in radiation.

The Ministry of Internal Affairs and Communications (MIC) took about 1 year from January 2008 to carry out an administrative evaluation and monitoring of overall administration of nuclear disaster prevention aimed at improving related ministries and agents. Of the 22 off-site centers nationwide, 13 of them were subjected to administrative evaluation. Among the 13 evaluated, 7 of them were located within the Emergency Planning Zone (EPZ: area within 10 km of the NPP).

MIC put the evaluation results into the report "Recommendations Based on Administrative Evaluation and Monitoring of Administration of Nuclear Disaster Prevention (second Stage)" [7] and pointed out that five centers including Fukushima did not have an air-ventilating system that lowers radiation exposure and made a recommendation to correct the situation.

NISA decided to improve the air tightness of the off-site center building and better manage the building entry system; however, they ignored the recommendation by not taking actions like installing an air-filter. NISA did not give sufficient thought to preparing for a large-scale nuclear power disaster.

There were also issues with the NISA staff dispatched to the site. When the earthquake hit, there were seven safety inspectors and one NISA main office staff at Fukushima-1. After the accident, three of the eight moved to the off-site center, and five remained within Fukushima-1. These five, however, moved to the off-site center before dawn on March 12th because the radioactive levels went up in the area.

Later, the local response headquarters was concerned about having no NISA staff on site at Fukushima-1 and decided to dispatch four, who arrived early in the morning of March 13th. These four, however, were not sufficiently active in information gathering such as stepping out of the seismic isolated building to confirm the water injection work. When the reactor building of Unit 3 blew up and the situation of Unit 2 worsened, thinking that they were at risk, they left Fukuhisma-1 at about 17:00 on March 14th without instructions from the local response headquarters.

Another important organization in nuclear regulation was the Nuclear Safety Commission. The commission audited and reviewed safety regulations by NISA and other regulating bodies as a third party and also specified evaluation standards for safety review by the regulating bodies. It also had the task of providing technical support to the central and local governments in case of nuclear power disasters.

The Nuclear Safety Commission, as we pointed out earlier, performed their tasks with their own assumptions that safety with Japanese nuclear facilities was sufficient, the possibility of severe accidents was negligibly small from an engineering perspective, and risks with nuclear reactor facilities were sufficiently small. The severe accident of Fukushima-1 revealed that these ideas were wrong.

3.3.3 Assessment of information management and public announcements

Given its wide impact, the manner in which related organizations supply information to the country and also to the outside world is extremely important, especially with a continuously changing situation such as a nuclear power disaster.

In such a situation, a one-way transfer of administrative and specialist judgments is called risk messaging. In the case of nuclear disasters, highly specialized technical information that is hard to illustrate and data about radioactivity are the subjects of information transfer. In these cases, one-way risk messaging can easily cause confusion and doubts among the population at large. When making information transfer, we shall first understand what it is that people, especially residents in the local area, want to know and then review how people accepted and understood the information.

Public announcements about Fukushima NPP accidents were, at first, made independently by (i) the Chief Cabinet Secretary, (ii) NISA, (iii) local response headquarters, (iv) Fukushima prefecture, and (v) TEPCO; however, on and after March 12th, all announcements were made after approval by the Prime Minister's Office, and from April 25th on, the government's and TEPCO's announcements were coordinated into the same announcements, and press releases were held at the General Headquarters.

The manner in which these organizations made information about the Fukushima-1 NPP accident public caused some confusion in the minds of local residents and the general population; especially the announcements about the spread of radioactivity, its prediction, state of the cores, and the critical situation of Unit 3.

As an example, the Chief Cabinet Secretary and others described radiation effects on the human body using the expression "no immediate effect on the human body." The phrase "no immediate" created fears among some that "there may be no effect now, but it may cause bad effects in the future" and was an unhelpful expression.

Another serious error by NISA was in replacing the councillor in charge of public announcements who, on March 12th, referred to a possible meltdown during a press conference. The abrupt replacement gave the impression to the public that NISA was concealing something.

Whatever the circumstances were, delays in providing information that required quick announcements, holding back press releases, and giving explanations that were vague and hard to understand interfered with proper judgments by the local residents and caused unnecessary anxiety and confusion.

The fundamental principle of public announcements is to quickly, precisely, and plainly explain the facts. Sticking to this even at a time of emergency is absolutely necessary to help local residents to make their own judgments and avoid causing anxiety and confusion among the people.

3.4 Local government: key responsibilities

3.4.1 Assessing the performance of local government in disaster management

Article 5 of the Nuclear Emergency Preparedness Act required local governments to "take actions necessary for the implementation of measures to prevent nuclear emergency, emergency response measures and measures for restoration from nuclear emergency." This article means that local governments are expected to, under the instructions and suggestions from the central government, prevent nuclear emergency, take emergency response measures and measures for restoration from nuclear emergency, as the administrative body that actually carry out these missions.

From the accident outbreak on March 11th to the following 12th, as the situations with Fukushima-1 NPP worsened, the central government kept expanding the evacuation and shelter zones while the overall status of the plant remained unclear. Evacuation orders by the central government failed to quickly reach all local governments of the areas and, furthermore, lacked preciseness and proper procedures.

Local governments only had the same level of information that was available from TVs and radios about the nuclear accident and had to lead residents to make the decision to evacuate, find destinations for them, and come up with actual ways of carrying out the evacuation. The situation led to massive confusions in the local area.

The actual records of initial evacuation shows (e.g., in case of the town of Namie) that the functions of the town hall and residents in the immediate vicinity of the plant were relocated to another area far from the plant but within the same town border. On March 15th, however, the town officials were told that the new area was also at risk, and they had to head for the city of Nihonmatsu. The route they took to Nihonmatsu was, however, later revealed to be along the direction of the radiation spread. Also, in the case of the town of Tomioka, residents first took shelter in the village of Kawauchi, but later together with the people of Kawauchi, had to move again to the city of Koriyama.

The evacuation of Futaba hospital

Fukushima prefecture, immediately after the Tohoku Area Pacific Offshore Earthquake, set up the Fukushima Prefecture Disaster Response Headquarters headed by the governor in the large conference room on the third floor of Fukushima Prefecture Hall next to the prefectural government building. After the outbreak of Fukushima-1 NPP accident, the Nuclear Power Group was set up within the headquarters to respond to it. The prefectural disaster prevention headquarters, in the middle of insufficient information and confusion with the unprecedented complex disaster, noted a number of problems. One such problem was the evacuation and rescue of inpatients at Futaba hospital.

Futaba hospital in the town of Ohkuma, next to Fukushima-1 NPP, received the order for evacuation in the early morning of March 12. The 209 patients that could walk on their own and all the hospital staff except the director boarded large chartered buses to evacuate Futaba hospital. At this point, however, about 130 patients confined to their beds, the hospital director, 98 members in the sister nursing house Deauville Futaba, and two of its staff were left at the hospital.

The prefectural disaster response headquarters learned about the people left at Futaba hospital and Deauville Futaba in the morning of March 13th when they received a rescue request from the off-site center. The prefectural disaster response headquarters sent a rescue and transportation request to the Japan Self-Defense Forces.

The Japan Self-Defense Forces rescue team arrived before dawn on March 14th, and started transporting the 98 in Deauville Futaba and 34 of the Futaba Hospital inpatients at around 10:30. After screening, the party arrived at Iwaki Koyo high school 8 h later at about 20:00 on the same day.

The high school expressed concern about accepting such ill patients without medical supplies and doctors. Iwaki Kaisei hospital, however, promised to dispatch doctors to the school, and Iwaki Koyo high school agreed to accept the patients. The patients started to be carried into the gym at about 21:30 on the same day, and eight of them were found to have passed away during the transportation.

At about 11:00 on March 15th, while radioactivity was high, the Japan Self-Defense Force started transporting the 47 left in Futaba hospital, and another Japan Self-Defense Force team rescued 7 at about 11:30. These 54 patients, after screening, headed toward Fukushima Medical University Hospital which was unable to admit them. At about 01:00 on March 16th, they arrived at Date Fureai Center. Two patients died during the transportation of this group.

Lastly, at about 00:30 on March 16th, rescue of the 35 that had remained in Futaba hospital started, but again 5 of them passed away during transportation.

This tragedy was caused by the following circumstances:

Firstly, Fukushima prefecture's disaster prevention plans had teams for resident evacuation, safety (responsible for arranging evacuation vehicles), and rescue (responsible for counting the number of remaining patients and securing destination hospitals). However, actual members of the teams had multiple assignments, and there was no managing group for overall control. This meant that there was no clear idea of where the evacuees were located until March 13th.

Secondly, the prefectural emergency response headquarters had learned that most of the inpatients at Futaba hospital were confined to bed. However, they had not shared that information within the headquarters. Then the transportation arrangement on March 14th supplied vehicles not suited for moving patients confined to bed.

Thirdly, the prefectural department in charge had already arranged for destination hospitals, aside from the prefectural disaster response headquarters, but the information had not been given to headquarters, and the transportation team ended up in a high school gym in a remote location.

Lastly, on the evening of March 14th, the director of Futaba hospital took shelter with policemen in a separate location to wait for the Self-Defense Force team to arrive. The information was relayed to the prefectural emergency response headquarters, but was not shared within the headquarters. The Self-Defense Force team failed to meet the director and policemen on March 15th, and the team of the second rescue mission that day never knew about the 35 remaining patients, who had to wait past midnight for rescue.

In all, the Fukushima emergency response headquarters had initial problems in clarifying responsibility assignments for each group and sharing information. The shortcomings led to insufficient administration in preventing the spread of the disaster.

Assessing the accident response by the prefecture

The Fukushima prefectural government made inadequate moves in distributing stable iodine tablets.

Stable iodine, when taken before exposure to radiation, prevents radioactive iodine accumulation in the thyroid gland, even if the material enters the body. The Nuclear Safety Commission compiled "About preparatory administration of stable iodine at times of nuclear disasters" [8], in which it wrote, "The disaster response headquarters shall make judgments for preparatory administration of stable iodine, in addition to preventive measures of taking shelter and evacuation" [8] and local towns that surround Fukushima-1 and -2 had prepared stable iodine tablets before the accident.

The town of Miharu, late at night on March 14, decided to distribute and administer stable iodine tablets to prepare the residents against radiation exposure. At around 13:00 on March 15th, the town made an announcement over the wireless town loud-speaker system and, under the supervision of the town pharmacist, distributed stable iodine tablets to about 95% of the residents. Learning about this distribution, a staff member at Fukushima Health and Welfare Department, in the late afternoon on the same day, ordered the town to stop the distribution and recall the tablets because the central government had not instructed the administration at that time. Miharu ignored the order and did not recall the tablets.

The administration of stable iodine tablets is formally subject to a decision by the disaster response headquarters of the central government; however, the move by Miharu was to protect the health of the local residents. Opposition by the prefecture staff for receiving instructions from the central government was inappropriate.

As in the preceding examples, the local government made a number of misjudgments as a result of confusion and lack of information (see Ref. [24] for comments on the local government response).

3.5 Conclusions: key lessons and improvements needed

A Nobel Prize economist, Herbert Simon, explained that humans have limits to rational behavior and named the concept "limited rationality" in his book *Administrative Behavior: A Study of Decision-Making Processes in Administrative Organizations* [9]. In other words, Simon pointed out that there is a limit to the intellect of humans, and we cannot predict all problems that will happen in the future. We can never overcome our limitations, and learning from troubles and accidents that did take place helps us expand our views and ranges of recognition.

In January 1995, the Great Hanshin-Awaji Earthquake occurred, killing thousands of people; this occurred 47 years after the Fukui Earthquake in 1948 that took 3769 lives. The Tomiichi Murayama Cabinet at the time of the 1995 quake made a number of mistakes.

Sixteen years later, with the Great East Japan Earthquake, the Self-Defense Force quickly entered the picture after learning from the Great Hanshin-Awaji Earthquake. The earthquake and tsunami with the Great East Japan Earthquake completely destroyed many lifelines of electricity, gas, and water; however, a number of business organizations made great efforts in helping the region, another lesson learned from the Great Hanshin-Awaji Earthquake.

The Fukushima-1 NPP accident, on the other hand, faced simultaneous problems with three of its nuclear reactors and progress with one nuclear reactor affected actions and responses with neighboring reactors. Up to that point, severe accident management in Japan never had to deal with severe simultaneous accidents with multiple nuclear reactors. No preparation had been made for such an event, and the actual responses gave rise to a number of problems and questions.

The International Nuclear and Radiological Event Scale (INES) ranks problems with nuclear power facilities in terms of safety in eight levels of 0-7. Levels 1 through 3 are categorized as incidents, and 4-7, accidents. An incident is an event that threatens or possibly threatens safety. The distinction between an incident and an accident is in the size of the effect. When the effect reaches outside of the facility and at least one person dies from the effect of radiation, the event is an accident.

Up to the point of Fukushima-1 NPP accident, the most severe incident in Japan was the 1999 Tokai JCO criticality accident at its nuclear fuel processing plant (two workers died from fatal radiation exposure). The second worst was at the Power Reactor and Nuclear Fuel Development Corporation (later merged into Japan Atomic Energy Agency in 2005) fire and explosion accident in 1997 at its Tokai asphalt caking facilities (37 workers were exposed to radiation). The former was evaluated at Level 4, and the latter at Level 3. Both accidents took place in nuclear power-related facilities and not NPPs. NPPs in Japan had only experienced Level 2 or less severe troubles before, not to mention none that released a large amount of radioactivity.

The Three Mile Island and Chernobyl accidents showed that severe accidents that exceed the design standards could happen with NPPs. The Fukushima-1 accident reminded us of the importance of learning from past accidents.

References

[1] IAEA, Safety of Nuclear Power Plants: Design Specific Safety Requirements, Series No. SSR-2/1.
[2] Act on Special Measures concerning Nuclear Emergency Preparedness (Nuclear Emergency Preparedness Act, 1999).
[3] Accident Management: Measures against Severe Accidents at Light Water Nuclear Reactor Facilities for Power Generation (Common Problem Round-Table Group within the Special Committee on Nuclear Reactor Safety Standards and Guides) in Japanese: http://www.mext.go.jp/b_menu/hakusho/nc/t19920528001/t19920528001.html, May 1992. Viewed 20.08.13.
[4] Future Processes of Accident Management (METI). http://www.nsr.go.jp/archive/nisa/shingikai/800/34/003/sankou4-4-1.pdf. Viewed 20.08.13.
[5] Interim report IV-4-(2)-b (pp. 416–418 in the Japanese version).
[6] Regulatory Guide for Reviewing Seismic Design of Nuclear Power Reactor Facilities NSC Decision No. 2006-D59, Appendix 1, NSC, 2006. http://www.nsr.go.jp/archive/nsc/NSCenglish/documents/decisions/2006/2006_59_1.pdf. Viewed 20.08.13.
[7] Recommendations Based on Administrative Evaluation and Monitoring of Administration of Nuclear Disaster Prevention (2nd Stage) (Ministry of Internal Affairs and Communications, Feb., 2009).
[8] About preparatory administration of stable iodine at times of nuclear disasters, 2002, NSC.
[9] Administrative Behavior: A Study of Decision-Making Processes in Administrative Organizations, Herbert Simon, THE FREE PRESS.
[10] Interim report: VII-4. (pp. 472–476 in the Japanese, pp. 570–575 in the English version).
[11] Interim report: VII-6-(2) (pp. 492–495 in the Japanese, pp. 590–593 in the English version).
[12] Final report: VI-1-(6) (pp. 401–406 in the Japanese, pp. 471–476 in the English version).
[13] Interim report: V-3-(1),(2) (pp. 263–277 in the Japanese, pp. 302–318 in the English version).
[14] Interim report: V-8,9 (pp. 349–360 in the Japanese, pp. 410–423 in the English version).
[15] Interim report: VI-7-(2) (pp. 455–460 in the Japanese, pp. 544–551 in the English version).
[16] Interim report: VII-3 (pp. 467–471 in the Japanese, pp. 565–570 in the English version).
[17] Investigation Committee on the Accident at the Fukushima Nuclear Power Station of Tokyo Electric Power Company, Interim report. http://www.cas.go.jp/jp/seisaku/icanps/eng/interim-report.html, 2011. Viewed 20.08.13.
[18] Investigation Committee on the Accident at the Fukushima Nuclear Power Station of Tokyo Electric Power Company, Final report. http://www.cas.go.jp/jp/seisaku/icanps/eng/final-report.html, 2012. Viewed 20.08.13.
[19] Interim report: V-3-(1),(2) (pp. 263–277 in the Japanese, pp. 302–318 in the English version).
[20] Interim report: V-8,9 (pp. 349–360 in the Japanese, pp. 410–423 in the English version).
[21] Interim report: VII-5-(1),(2),(3) (pp. 477–481 in the Japanese, pp. 575–580 in the English version).
[22] NSCRG: L-DS-I.0, Regulatory guide for reviewing safety design of light water nuclear power reactor facilities, NSC. http://www.nsr.go.jp/archive/nsc/NSCenglish/guides/lwr/L-DS-I_0.pdf, 1990. Viewed 20.08.13.
[23] Interim report: VI-4.-(1)-b. (pp. 410–414 in the Japanese, pp. 485–489 in the English version).
[24] Interim report V-4.-(5)-e (pp. 306–309 in the Japanese, pp. 356–361 in the English version).
[25] About "Review Guide for Safety Design of Light Water Nuclear Power Reactor Facilities" (in Japanese), http://www.aec.go.jp/jicst/NC/about/ugoki/geppou/V22/N06/197702V22N06.html, Japan Atomic Energy Commission, June 1977.

Appendix Government offices and organizations

Government
Cabinet
Prime Minister
(Kan, Naoto)
Special Advisor to the Prime Minister
(Hosono, Goshi)
Cabinet Secretariat
Chief Cabinet Secretary
(Edano, Yukio)
Deputy Chief Cabinet Secretary for Crisis Management
(Ito, Tetsuro)
(General Administrative Agency, a.k.a Prime Minister's Office [–2001])
(Science and Technology Agency [–2001])
(Nuclear Safety Bureau [–2001])
Headquarters for Earthquake Research Promotion (transferred to MEXT in 2001)
Cabinet Office
(NERHQ: Nuclear Emergency Response
(Emergency Disaster Response Headquarters [set up earlier against the earthquake])
Japan Atomic Energy Commission
(NSC: Nuclear Safety Commission)
(Chair)
(Madarame, Haruki)
(Special Committee on Nuclear Reactor Safety Standards and Guides)
(Investigation Committee on the Accident at the Fukushima Nuclear Power
(Chair)
(Hatamura, Yotaro)
(Technical Consultant)
(Abe, Seiji)
(Fuchigami, Masao)
Central Disaster Management Council
Committee for Policy Planning on Disaster Management
Reconstruction Agency [Feb. 2012–]

Continued

Appendix Continued

(MITI: Ministry of International Trade and Industry [–2001])

(Environment Protection and Industrial Location Bureau [–2001])

METI: Ministry of Economy , Trade and Industry

Minister

Industrial Science and Technology Policy and Environment Bureau

Agency for Natural Resources and Energy

(NISA: Nuclear and Industrial Safety Agency)

(Director-General)

(Hirose, Kenkichi)

(Terasaka, Nobuaki)

(Vice Director-General)

(Hiraoka, Eiji)

MEXT: Ministry of Education, Culture, Sports, Science and Technology

NUSTEC: Nuclear Safety Technology Center (operates SPEEDI)

MIC: Ministry of Internal Affairs and Communications

Fire and Disaster Management Agency

MLIT: Ministry of Land, Infrastructure, Transport and Tourism

Japan Meteorological Agency

Tohoku Regional Bureau

MOE: Ministry of the Environment

NRA: Nuclear Regulation Authority[Sept.2012–]

Nuclear Regulation Agency

MAFF: Ministry of Agriculture, Forestry and Fisheries

MOD: Ministry of Defense

Self-Defense Force

Tohoku Regional Team

(Ministry of Transport [–2001])

(Ministry of Construction [–2001])

Continued

Appendix Continued

Offsite Center →Fukushima Prefecture Hall

- (Local NERHQ: Local Nuclear Emergency Response Headquarters [Article17, Section 9])
- Joint Council for Nuclear Emergency Response
- Nuclear Disaster Prevention Staff(TEPCO)

Local Government

- Fukushima Prefecture
- Tokyo Prefecture
  - Tokyo Fire Department
  - Metropolitan Police Department
    - Mobile Rescue Unit

TEPCO

- Headquarters
  - President
    - (Shimizu, Masataka)
  - Vice president
    - (Mutoh, Sakae)
  - Fellow
    - (Takekuro, Ichiro)
  - (Main office Emergency Response Headquarters)
- Fukushima 1
  - Plant Manager
    - (Yoshida, Masao)
  - (Plant Emergency Response Headquarters in Seismic Isolated Building)
    - (Electricity Team)
  - Recovery Team
    - Seismic Isolated Building
  - Subcontractor
    - for example, (Nanmei Kosan) now Toden Fuel

Others

- Futaba Hospital
- JSCE: Japan Society of Civil Engineers
  - Nuclear Engineering Committee
    - Tsunami Evaluation Workgroup
- JSME: Japan Society of Mechanical Engineers
- AESJ: Atomic Energy Society of Japan

Although JSME and AESJ do not appear in this book, they play important roles in the nuclear power industry from the academic side.

Continued

Appendix Continued

- Overseas
  - Global
    - United Nations
      - IAEA: International Atomic Energy Agency
      - WHO: World Health Organization
      - UNSCEAR: United Nations Scientific Committee on the Effects of Atomic Radiation
  - OECD/NEA: OECD Nuclear Energy Agency
    - Europe
      - EURATOM: European Atomic Energy Community
      - France
        - ASN: French Nuclear Safety Authority
          - President
            - (Lacoste, André-Claude)
      - Sweden
        - Swedish National Board of Health and Welfare
          - Director General
            - Holm, Lars-Erick
    - USA
      - US Armed Forces
      - NRC: Nuclear Regulatory Commission
      - GE

Organizations in parentheses no longer exist. Personal names in parentheses indicate the person is no longer in the position.
If the gray text has no ending year in square brackets, it ended after the earthquake.
Many organizations were transferred to new ones in the 2001 reform.

The response of the plant owner/ operator (TEPCO) to the Fukushima nuclear power plant accident

4

S. Abe
Kansai University, Osaka, Japan

4.1 Introduction

Tokyo Electric Power Company (TEPCO), the largest utility company in Japan and one of the largest electric utility companies in the world, was the first in line of responsibility to ensure the safety of the public. The company clearly struggled to manage the disaster.

This chapter reviews this utility giant to find where the weakness in its response came from. The direct actions and physical phenomena were described in Chapter 2, whereas in this chapter, we will take a second look at the actions taken and identify their root causes.

4.2 Accident management by Fukushima's nuclear plant owner and operator, TEPCO

4.2.1 TEPCO's use of accident management guidelines

As mentioned in Chapter 3, the Nuclear Safety Commission, in May 1992, set out "Accident Management: Measures against Severe Accidents at Light Water Nuclear Reactor Facilities for Power Generation" [1] and recommended that utility companies promote their own accident management. MITI also guided the utility companies to promote accident management as their own project.

Given these directions, TEPCO took about 10 years to establish its accident management for Fukushima-1, Fukushima-2, and Kashiwazaki-Kariwa and submitted their "Accident Management Development Report" [2] and "Accident Management Development Effectiveness Evaluation Report" [3] in May 2002 to METI (MITI up to January 5, 2001).

The accident management development by TEPCO at this time included the following four arrangements: (i) developing accident management measures like water injection functions to the reactor core and Containment Vessel (CV), (ii) assigning groups and organizational structures for actual accident management, (iii) building procedural documents like operational procedures at times of accidents and an accident

The 2011 Fukushima Nuclear Power Plant Accident. http://dx.doi.org/10.1016/B978-0-08-100118-9.00004-8

management guide, and (iv) education of employees assigned to accident management groups.

As noted in the previous chapter, causes of severe accidents were mainly limited to internal factors, and thus, accident management against external factors like natural disasters was not sufficiently covered in the scope of the accident management described here.

TEPCO settled with the accident management completed in 2002, and from then on, it concentrated on "lateral dissemination" of countermeasures learned from accidents and findings at domestic and overseas nuclear reactor facilities. For example, after the July 2007 Chuetsu Offshore Earthquake, Kashiwazaki-Kariwa suffered damage to the office building and a fire in the transformer area, and the "lateral dissemination" led to the stationing of two chemical fire engines and a tank-mounted fire truck at Fukushima-1 by February 2008. Also in 2010, several new fire reservoirs were placed, water inlets were newly installed on the turbine building for each unit to reach the cooling systems, and the emergency response headquarters were transferred to the seismic isolated building from the administration area of the main building.

The seismic isolated building was equipped with emergency power generators and protection against radiation. That building played an important role in this emergency response at Fukushima-1. If the seismic isolated building had not been available, accident response at Fukushima-1 would have been much more difficult, and this point was commendable.

As seen earlier, TEPCO carried out lateral dissemination from 2003 on, but did not push for stronger accident management, including measures against external factors.

Japan is an earthquake-prone country. Natural disasters like earthquakes and tsunamis have to be considered as potential causes of severe accidents. TEPCO, however, did not consider ways of preventing damage to the reactor core in the case of severe accidents caused by natural disasters that exceed the design assumptions. TEPCO's concept of preparation against natural disasters was to assume certain levels of natural disasters like earthquakes or tsunami and that it was sufficient to design the nuclear reactor facilities in accordance with the Guidelines for Safety Design Reviews and Guidelines for Aseismic Design Review by the Nuclear Safety Commission. TEPCO (as well as other utility companies in Japan) did not prepare adequate countermeasures against severe accidents with its existing nuclear reactor facilities. Instead, they ran a series of aseismic back-checks (evaluation of aseismic safety) to see if the facilities could withstand natural disasters of the set magnitudes, and if locations with insufficient resistance were found, they were modified as countermeasures.

The Fukushima NPP accident suffered severe damage to three nuclear reactor cores at one time. Submergence of the facilities led to station blackout (SBO), for which the station was unprepared. There was an underestimation of risk with regard to exceptional external events. TEPCO thus left the preparations against severe accidents in an incomplete state.

4.2.2 TEPCO's measures for accident management

What accident management measures were actually in place at Fukushima-1? We will review its (i) measures against loss of power, (ii) preparation of fresh- and seawater injection from fire engines, and (iii) communication tools in emergency situations before the accident took place.

Measures against loss of power

TEPCO's measures against loss of power had been built with the assumption that at least one of the neighboring nuclear reactor facilities would be intact. In other words, no preparation had been discussed in case there was no power supply available from a neighboring reactor facility due to simultaneous failure of multiple plants for some reason.

Also, emergency diesel generators had been added since the time of licensing. However, no switchboard had been additionally installed for multiplicity (i.e., there was no preparation against an extended SBO with all external and internal power sources unavailable).

No plans had been made for these scenarios, no documentation had been prepared for procedures to recover instrumentation, electrical power, CV vent, and Safety Relief Valve (SRV) operation in such severe circumstances, and there was no training in these procedures. The Fukushima-1 site had no battery, air compressor, electrical power vehicle, or electrical cable necessary for such procedures.

Preparation of fresh- and seawater injection from fire engines

The lateral dissemination of the fire at Kashiwazaki-Kariwa caused by Chuetsu Offshore Earthquake led to stationing of fire engines at Fukushima-1 as well. Though the effect of water injection from fire engines was shared among some in the company, it was not an established procedure in the accident management measures.

Although seawater injection was seen as an alternative for the worst situations, it was judged an unlikely scenario and was never discussed seriously. Alternative water injection from fire engines also was never seriously considered, and which team or group from the response headquarters would carry out the procedure was not clear.

Thus, at 17:12 on March 11th, when the Plant Manager Yoshida ordered alternative water injection from fire engines, leaders and members of all the teams thought the order was intended for someone else, and the preparation took as long as 2 hours.

Also, continuous alternative water injection from fire engines meant securing a water source, and ultimately it would call for seawater injection. However, no preparation or arrangement for seawater injection had been made. Therefore, on March 12th when it became a reality, the piping for injection took significant time to arrange.

Failure of the emergency communication system to function

At a time of emergency, not only at Fukushima-1 but also at every other plant, communication between the field, the plant response headquarters, and the central control room is important. Thus, it is indispensable to prepare means of communication that are available during emergencies.

Up to the time of the accident, Fukushima-1 had been using PHS as the means of communication, and it was expected to work at a time of emergency as well. The reality was, however, that the backup battery in the PHS base station only lasted for 3 hours, and in the evening of March 11th, the PHS units became inoperable one after another.

Then the workers on-site, at plant response headquarters, and in the central control room lost their means of communication. They tried wireless communicators such as walkie-talkies, but they had limitations in the reach of wireless signals. Communication during the initial stage of the emergency response was very difficult.

TEPCO had set a minimum battery life of 1 hour for transmission and switching equipment for PHS-related machines in the nuclear facilities. This setting was based on the idea that AC power would recover within 1 hour from SBO. Again, an extended SBO had not been considered.

Incomplete accident management plans were not unique to Fukushima-1; other nuclear power stations in Japan had the same problem until the Fukushima NPP accident occurred.

The Federation of Electric Power Companies of Japan, consisting of all electricity companies in Japan, had on its homepage a frequently asked question of "Why is accident management needed against severe accidents that will not take place?" (January 20, 2013).

The answer to this question stated, "A nuclear power plant goes through strict safety assurance measures from design, construction, to operation management based on the concept of multi-layered protection, and safety is at a sufficiently high level. Development of accident management is a voluntary measure by the electricity companies from the standpoint that continuous efforts bring benefits to further improve safety, not withstanding the sufficiently high safety of nuclear power plants".

Accident management would turn out to be the most powerful card to play in responding to a severe accident. The reality, however, with Japanese electric companies was that they shared a common misconception that "nuclear power plants are sufficiently safe" and accident management was given the position only of a "precautionary measure."

4.3 TEPCO's risk assessment and safety planning documentation at the Fukushima plant

In July 1993, the Southwest Hokkaido Offshore Earthquake and tsunami caused great damage to Okushiri Island and other areas. This tsunami triggered a review of tsunami preparation in Japan. In March 1997, the Ministry of Construction, the Ministry of Transport, the Ministry of Agriculture, Forestry, and Fisheries, and other related ministries published the "Research report on earthquake tidal wave disaster prevention plan for Pacific coast division" [4] to show new concepts of tsunami disaster prevention and methods for evaluation.

Given this publication, the electricity companies started an electrical power joint research of "Research on enhancing technology of tidal wave evaluation." In 1999, the Japan Society of Civil Engineers, Nuclear Engineering Committee formed the Tsunami Evaluation Workgroup to discuss systematized and standardized technology for safety evaluation of nuclear power facilities against tsunami.

The workgroup in February 2002, documented the technology of tsunami evaluation into "Tsunami Assessment Method for Nuclear Power Plants in Japan" [5], including results from the electrical power joint research group. The study made some changes (e.g., the Fukushima-1 expected tsunami height was lifted from 3.1 to 5.7 m). The method for estimating the tsunami height was valid; however, it had the following problems.

During the workgroup discussions, some pointed out the possibility of a tsunami that exceeded the tsunami height expectation; however, the Tsunami Assessment Method concluded that no tsunami would exceed the calculated maximum tsunami height. The assessment could have listed the applicable extent of the technology applied and its limitations. Such discussions could have led to cautions against the problems with tsunami in revising the Review Guide for Safety Design; nonetheless, such elaboration was not in the document.

The Tsunami Assessment Method estimated the highest tsunami by evaluating past tsunami waves with somewhat reliable traces of their heights. Therefore, it was based only on tsunami waves that attacked the coasts in the previous 300-400 years. Tsunami with long cycles of 500-1000 years were neglected if there were no documents or data available. The "Research Report on Earthquake Tidal Wave Disaster Prevention Plan for Pacific Coast Division" was the background data for Tsunami Assessment Method, and it was written as countermeasures to take for preparation against tsunami. The Tsunami Assessment Method, on the other hand, was the technology for estimating the tsunami height, and it was not intended to discuss what measures to take given the calculated height.

The 1995 Great Hanshin-Awaji Earthquake led to the enactment of the "Act on Special Measures for Earthquake Disaster Prevention" [6] in the same year. The act called for establishing The Headquarters for Earthquake Research Promotion in the General Administrative Agency of the Cabinet (transferred to Ministry of Education, Culture, Sports, Science and Technology (MEXT) in 2001). The Headquarters for Earthquake Research Promotion published, in July 2002, "Long Term Assessment of Seismic Activities from Sanriku Offshore to Boso Offshore" [7], stated, "An earthquake similar to the 1896 Meiji Sanriku Earthquake can occur anywhere from the northern area of Sanriku offshore to along the ocean trench in the Boso offshore."

The Nuclear Safety Commission revised the guidelines for safety reviews on aseismic safety in September 2006 [8]. The Nuclear and Industrial Safety Agency (NISA), in response in 2007, requested the electricity companies to reevaluate the aseismic capabilities, including tsunami preparedness assessment, of nuclear reactor facilities in operation and under construction at the time. During the back-check of

Fukushima-1 and -2 in response to this request, a concern was raised about how to deal with the statement by the Headquarters for Earthquake Research Promotion about a possible earthquake equivalent to the Meiji Sanriku Earthquake in the area from north of Sanriku offshore to Boso offshore.

TEPCO, in 2008, then reassessed its risk associated with tsunami to evaluate whether this statement would overturn the Fukushima-1 safety evaluation based on the 2002 Tsunami Assessment Method. The study concluded an estimated tsunami height of above 10 m at Fukushima-1. TEPCO also calculated the wave height based on the source model from the Jyogan tsunami described in the so-called Satake Paper [9]. This study returned a wave height of above 9 m.

TEPCO, however, took the former result as a hypothetical number from placing the Sanriku offshore source model to Fukushima offshore, and concluded that the latter was not based on a sound source model and that the statement by the Headquarters for Earthquake Research Promotion lacked sufficient evidence. As a result, they did not review and enhance the tsunami preparation at Fukushima-1.

When a commercial business makes such a decision, NISA, rightfully as the regulating body for nuclear safety, should have evaluated whether the decision was reasonable or not, but instead, NISA took no action and did not oppose TEPCO's decision.

As discussed here, there was a chance to review the tsunami preparation in 2008; however, the chance was not taken. There was an underestimate of the risk of tsunami-induced severe accidents by TEPCO, NISA, and the Nuclear Safety Commission.

4.4 Assessing the postaccident responses of TEPCO

4.4.1 Division of roles between the central headquarters and plant

TEPCO, following Article 10, Paragraph 1 of the "Act on Special Measures Concerning Nuclear Emergency Preparedness," established a rule that when an event specified in the paragraph occurred, and the nuclear emergency preparedness manager or the plant manager declared the first degree state of emergency, applicable nuclear power plants and the central headquarters should establish their emergency response headquarters.

At 15:42 on March 11th, Fukushima-1 Plant Manager Yoshida judged that an event specified in the "Act on Special Measures Concerning Nuclear Emergency Preparedness (Nuclear Emergency Preparedness Act)," Article 10, Paragraph 1 had taken place and notified applicable ministries and agents, local governments, and the TEPCO main office. The notification was responded to by settings of emergency response headquarters in the TEPCO main office and at Fukushima-1. At that time, the emergency disaster response headquarters had already been placed an hour earlier in the seismic isolation building with the Great East Japan Earthquake attack (Figure 4.1). The emergency response headquarters in Fukushima-1 merged the one for earthquake response within to lead the overall disaster response management.

Figure 4.1 Seismic Isolated Building. Built in July 2010, only 8 months after the Chuetsu Offshore Earthquake as a result of the lessons learned from it. This facility, with an independent power generator, functioned quite effectively during the accident.

Figure 4.2 shows the relationship between the local plant headquarters and the main office headquarters. The local Plant Emergency Response Headquarters were at the frontline of emergency responses, and the Main Office Emergency Response Headquarters supported the local headquarters by confirming and giving authorization for important procedures.

Emergency responses in Fukushima-1 NPP were carried out while all lights were lost and the rubble of equipment and facilities were scattered everywhere except in the seismic isolated building. In the late afternoon on March 11th, to worsen the situation, damage to the nuclear reactor had started to release radiation near the reactor. Field workers of TEPCO and its contractors had to work in pitch darkness with high levels of radiation. The hardship they had to tolerate is beyond what we can imagine.

Although the circumstances of the work area were difficult, TEPCO's response to the emergency showed weaknesses. In the following paragraphs we will discuss the failure to correctly assess the operational status of the Unit 1 Isolation Condenser (IC) and weaknesses in handling Unit 3 alternative water injection.

4.4.2 Misjudgment of Unit 1 isolation condenser (IC) status

IC, as discussed in detail in Chapter 2, is an emergency cooling system that cools the Reactor Pressure Vessel (RPV) steam in the condenser tank and feeds the condensed water back to the RPV by gravity without the use of a pump. Its continuous operation cools the reactor core. Fukushima-1 had two IC systems.

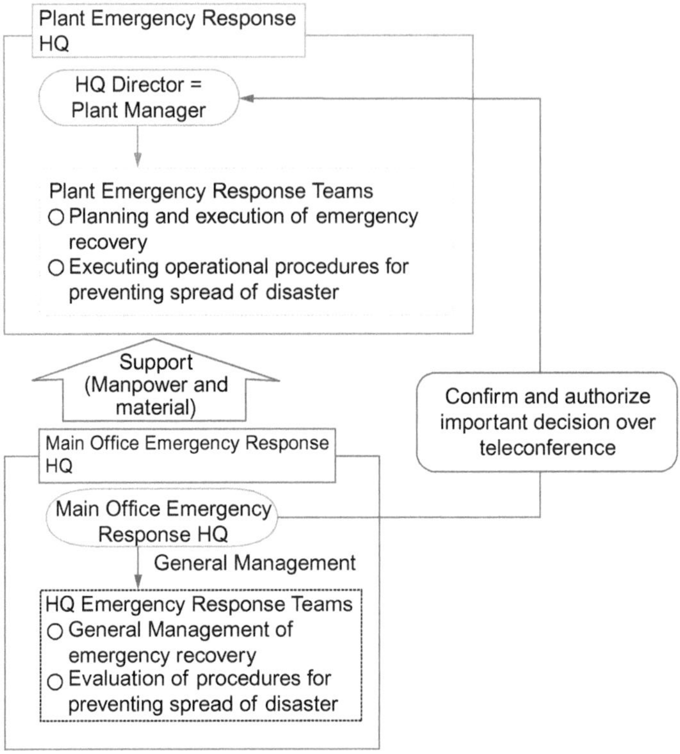

Figure 4.2 Organization of Emergency Response at TEPCO. [10]

If these ICs had operated normally, as we pointed out in Chapter 2, the core damage process of Unit 1 would have been different. The ICs, in reality, were not operating properly, and the Emergency Response Headquarters had mistakenly judged they were in proper operation.

The biggest reason for the misjudgment was, including the main office, TEPCO engineers' lack of knowledge about the basic operation of ICs. These IC systems were designed so their logic closes the IC isolation valves immediately after loss of DC power for a fail-safe purpose. None of the staff at the plant or the main office emergency response headquarters realized this.

Between 16:42 and about 16:56 on March 11th, there were symptoms of trouble with the ICs such as lowered reactor water level or high levels of radiation near the Unit 1 reactor building. The engineers should have realized the high likelihood of IC trouble with these phenomena. Hardly anyone at the response headquarters realized the trouble, and thus proper instructions were not given and the right actions were not taken in the field.

The operator in charge at 18:18 on the same day began to be concerned that the IC isolation valves might not be responding when he tried to open valves 2A and 3A after the control panel display was recovered. He then consulted with the plant response headquarters, but because they did not understand the IC basics, headquarters did not change their view on the IC operation.

The hearing by our investigation team of the Fukushima-1 engineers found that no one in the plant had recent experience with IC operation, and the operators had only heard of such experience by word of mouth. Testimony during the hearing claimed training had been given about functions and operations of the IC, but none in the case of loss of DC power.

The plant response headquarters was at the front end of accident response, and the main office response headquarters' role was to support the plant. No one in these two headquarters had a complete understanding of the IC mechanism and function. The operators in charge were not familiar with its operation, either.

Cooling the reactor to prevent core damage must have the highest priority at the time of an accident. In the authors' opinion a company with such poor understanding of the functions and handling of IC, that was most relied on at a time of an accident, lacked the qualifications to run a nuclear power plant.

The functional failure of IC called for a quick alternative water injection. This injection could not have been made without lowering the RPV pressure. At about 0:00 on March 12th, instructions were given to prepare to drop the pressure; however, the work did not start until about 14:00 on the same day.

A long delay in dropping the RPV pressure and injecting water meant a delay in the core cooling. The misjudgment about the IC operation was the biggest cause for this delay.

SBO is a seriously severe situation, and no matter what the circumstances were, measures to cool the core had the highest priority, However, the plant and main office response headquarters failed to recognize the IC operation status and thus, did not rush in creating a method of alternative water injection. In addition, instructions to vent the CV were also delayed. After all, misjudgment about the IC operation triggered the chain of delays in responding to the Unit 1 accident.

Core cooling of Unit 1 failed in the way described above, and in the afternoon of March 12th, a hydrogen explosion blew up the top floor of the reactor building. The explosion broke the temporary cables that had been laid out for electric recovery of Units 2 and 3, further delaying the work on these two other reactors. The misjudgment of IC operation, thus, not only affected the Unit 1 accident response, but also caused a negative impact on the responses for Units 2 and 3.

4.4.3 Mishandling of Unit 3 alternative water injection

After the Unit 1 reactor building exploded at about 15:36 on March 12th, continuous core cooling of each unit was recognized as the most critical process. The mishandling of alternative water injection for Unit 3 happened under such circumstances.

When water injection with a certain method fails, it is absolutely necessary to immediately switch to another method without any delay. At about 02:42 on March 13th, without an alternative method prepared, operators and others in the field manually turned off the high pressure core injection (HPCI).

Furthermore, a measure against discharged batteries had not been in place, the drained batteries could not restart the HPCI, and the pressure reduction for alternative water injection failed as well. Water injection to the reactor core was, thus, suspended for over 6 hours. To make matters worse, the trouble was not immediately reported to the senior staff. Alternative water injection did not start until about 09:25 on the same day, and the Unit 3 core was damaged.

The decision to stop the HPCI only with the Unit 3 staff at the time and the plant response headquarters power generation team without requesting authorization from a senior member of staff was questionable in terms of a response in an emergency situation. In addition, if the information about manual stoppage of HPCI had been fully shared within the plant response headquarters, there might had been a quicker remedy to the operator's misjudgment of manually stopping the HPCI without an alternative method for water injection.

4.4.4 Consequences of the inaccurate assessment of Units 1 and 3

As Chapter 2 discussed in detail, hydrogen explosions took place in the reactor building of Unit 1 at about 15:36 on March 12th, and in Unit 3 at about 11:01 on March 14th. The two buildings suffered great damage. These explosions were caused by damage with the fuel rods inside the reactor core. The Zircaloy that covered the nuclear fuel reacted with water to produce hydrogen. The hydrogen then made its way out from the RPV, through the CV to accumulate inside the reactor building.

The question is if the response headquarters had properly acknowledged the status of Units 1 and 3 and had dropped the pressure earlier to inject water, could the core damage have been prevented to avoid hydrogen generation and thus the explosions?

To precisely answer this extremely important question, further detailed investigation is needed about the core and possible water injection, and thus, finding an answer at this point is difficult. We can say at least that if earlier depressurization had been made for alternative water injection from fire engines, the core damage would have proceeded more slowly, and less radioactivity would have been released.

4.4.5 Faults by the local response headquarters and main office response headquarters

The May 2002 "Accident Management Development Report [2]" describes the basics of the Fukushima-1 emergency response. This report states, "For complex events, technical evaluation is highly important and vast information is required for understanding the accident status and deciding which accident management measure to take. The supporting organizations, therefore, perform these technical evaluations to support the decision making." In other words, the teams for information, engineering, safety, recovery, and power generation were expected to gather all necessary information to carry out the technical evaluation and provide suggestions and instructions to the chief on duty.

Let us take the event we explained earlier to describe this idea. When the supporting organization received information about Unit 1 IC and intended to cool the core in isolation, it would properly evaluate the IC status and in case such information was not available would call the chief on duty to actively collect the information. Unfortunately, the reality proceeded differently, with the headquarters failing to perform such a role and without correcting the misunderstanding of the IC status.

The teams of the main office response headquarters were also expected to gather important information from teleconferences and so to evaluate the information from viewpoints different from those of the plant response headquarters and support the plant's decision making. Each team in the main office did not provide effective advice to the plant response headquarters. Inability to provide effective advice and instructions when faced with the serious problem of delay in core cooling showed weaknesses in accident management for Fukushima-1.

When all AC power was lost, power supply from batteries was also headed toward drainage. When over a day had passed from the loss of all AC power before dawn on March 13th, engineers working on Fukushima-1 should have been concerned with the battery drainage for operating HPCI and the reactor core isolation cooling system (RCIC) of Unit 3. If they had, the plant response headquarters could have worked earlier on securing alternative water injection from fire engines instead of relying on HPCI running at the time but at risk of losing the function once it was stopped.

Also, before dawn on March 12th, with the rubble already removed, the fire engines that had been parked near Units 5 and 6 were available, and there was time for collecting batteries for activating SRVs for RPV depressurization.

The plant response headquarters at the time, however, had only discussed and prepared alternative water injection from the standby liquid control system, which is a mid- to long-term response, and did not start planning for alternative water injection using fire engines until hearing about the trouble of manually stopping the Unit 3 HPCI from the operator on duty. For Unit 2, Plant Manager Yoshida had given instructions to prepare alternative water injection around noon on March 13th before its RCIC stoppage; thus, if the reactor status was well understood, similar responses could have been taken for Unit 3 as well. The plant response headquarters did not recognize the urgent need to prepare alternative water injection for Unit 3.

It is understandable that the plant response headquarters had difficulty in facing an unprecedented emergency situation. The response headquarters in the main office was not effective in supporting the plant response headquarters.

4.5 Why procedures were inadequate: organizational problems within TEPCO

4.5.1 Weaknesses in emergency response capacity

TEPCO had been educating and training those in nuclear power generation to the level required by law before the Fukushima-1 NPP accident. The investigation committee interviewed a number of TEPCO employees, and those in the nuclear power department had ample knowledge of nuclear power technology at a level equivalent to plant manufacturers.

However, when we reviewed the responses by the employees to this Fukushima NPP accident, in many cases their knowledge was not used to its full extent. A typical example was the misjudging of the Unit 1 IC operational status, and a similar confusion can be noted about the reactor water level gauge.

During the accident, when the reactor water level gauge indication stood still for a long time, no one in the main office or on-site at Fukushima-1 voiced the possibility that the actual water level could be lower than the reactor side piping tap. Records show, however, that there was one worker who pointed out the possibility of a higher indication of the reactor water level but a lowered water level in the reference condensate pot. Nevertheless, no evaluation or discussion was made about such a possibility that could keep the reactor water level indication unchanged.

Also, the employees had plenty of knowledge about the containment atmospheric monitoring system (CAMS) and how to apply its readings to accident management; however, during and after the accident, no one tried to evaluate the plant status by estimating the RPV and CV integrity from the CAMS readings, and they simply followed the manual to calculate the core damage rate and reported the results to NISA.

It is understandable that the workers were under high tension in the middle of a difficult accident response; however, it can be argued that TEPCO showed real weakness in its capacity to respond to an emergency situation. This fact is due to the company not providing education and training to enhance such skills in coping with an emergency. The real cause was not the ability of each employee in the field.

When the authors further traced this problem of weakness in dealing with emergency situations, we concluded, finally, this was because companies such as TEPCO and even government officials underestimated the risk that a severe accident like a meltdown could happen with a Japanese nuclear power plant.

Employee qualities and competence in responding to severe accidents cannot be built in a matter of days, nor through classroom lectures. Ability to respond to accidents is gained not just from textbook knowledge, but also through the power of thinking of various possibilities from information at hand, making choices, judging what the best move is at the time, and executing the move.

TEPCO, as the first company in nuclear safety, should seriously review its education and training to enhance the quality and competence of every employee and contractor so they are ready to respond to accidents.

4.5.2 Problems with specialty-based sectionalism

TEPCO's disaster prevention plans and accident management guide called for teams such as power generation, recovery, engineering, and so on in the emergency disaster response headquarters. The scheme aimed at a systematic and integrated response to nuclear disasters. Each of the teams, however, tried to cover its own responsibility, but lacked managerial skills to see the total situation, find its role within this overall picture, and carry out the necessary supporting actions.

The corporate culture, similar to other electricity companies, was that TEPCO employees would refer to each other according to functions associated with their job, such as “operating guy,” “safety guy,” “electrical guy,” or “mechanical guy.” A few could gain key experience in a wide range of areas to ascend to a position at the director level; however, most employees in the nuclear power department were “some guy” that stayed in that area for years. Such workers have ample knowledge in their own fields, but do not have sufficient information to function in other fields, even those closely related to their own.

Once an organization is formed with such resources, the field of view for each employee is narrowed, and even when the organization appears trouble free at normal times, once an emergency situation breaks out, the weakness will be exposed. The organization cannot get a total and cross-sectional view of the situation, fails to decide what is important in prioritizing its actions, and thus ends up with a delayed start in important actions.

For example, Plant Manager Yoshida gave the order to evaluate water injection with fire engines soon after the accident on March 11th; however, it was not a documented procedure. Each functional team or group did not recognize its responsibility, and no actual evaluation took place until before dawn on March 12th. This was a typical example of the weakness described here.

Another case was the opening of SRVs. When electricity is available, it simply requires turning the switch on the control panel in the central control room. However, with the power lost, the recovery team had to connect the terminals in the back of the control panel to batteries with a total power of DC 120 V. There was confusion about whether the operator on duty or the recovery team was opening the Unit 2 SRVs on the evening of March 14th. This is another example of the weakness of sectionalism.

4.5.3 Lack of education or training for severe situations

As mentioned earlier, one of the reasons for the functional teams in the plant or main office response headquarters not performing their roles was sectionalism. In addition, poor performance of the functional teams was due to TEPCO not providing sufficient education and training to deal with total loss of AC power with multiple reactors.

TEPCO’s accident manual does not discuss multiple-unit SCRAM followed by days of SBO. These manuals assumed that AC power would recover within hours or, at most, a day. There was no description of the process of AC power recovery either. These manuals, at a glance, appear fairly detailed; however, they were incomplete with critical oversight.

The report TEPCO wrote in 2002, “Accident Management Development Report” described the loss of AC power with the phrase, “In an event that all AC power is lost, the phenomena will progress slowly with plenty of time margin”; however, no clarification was provided on why the phenomena would progress slowly. Preparing such incomplete procedures and training the employees with them can only cope with smaller events such as a local power loss.

Training was also insufficient. For example Fukushima-1 had, in late February 2011, a simulated training exercise assuming Nuclear Emergency Preparedness Act Article

10 notification. The postulation was loss of external power at a reactor caused by an earthquake, followed by the transformer break and failure to start the emergency generator. It was a state of loss of AC power. An assumption with this training, however, was that the emergency diesel generators would recover, and the simulation was on how to prevent catastrophic events from happening before the recovery of the generators. Nothing in the training had the severity of this accident, and no consideration was given to submergence of the switchboards or loss of internal power.

It has been claimed that loss of almost all power in Fukushima-1 caused by the earthquake and tsunami was beyond reasonable expectation. However, effective emergency planning should have taken more account of extreme events, as well as testing more effectively employees' ability to respond to emergency situations.

4.5.4 Excessive reliance on contractors

So far, in Fukushima-1 NPP, subcontractors had operated fire engines and heavy machinery; however, there had been no agreement that they would be handling them at times of emergency or abnormal situations. Their contract did not even mention a situation with this level of accident, where there is a high possibility of exposure to radiation (i.e., even the agreement with contractors ignored the possibility of severe accidents).

In the evening of March 11th, after the tsunami had subsided, the Fukushima-1 plant area was covered with rubble and wreckage of facilities that blocked people and vehicles. The plant wanted to remove these obstacles with a backhoe; however, there were no operators on site, and TEPCO had to make an urgent request to its contractor to dispatch operators. The contractor could have refused the request saying that it was beyond the scope of the agreement, but following their past practice, they did send operators to the site.

The next episode also shows TEPCO's heavy reliance on contractors: when the time came to inject water from fire engines, contractors like Nanmei Kosan had to run all the fire engines. None of the TEPCO employees knew how to operate them, and the water injection was delayed for that reason. Even though the hardware was ready, without knowing how to operate it, a quick initial response did not take place.

In the authors' opinion, TEPCO's excessive reliance on contractors had led to a situation where TEPCO employees lacked the hands-on skills to carry out the tasks for maintenance and emergency responses. The over-reliance of TEPCO on contractors and subsequent lack of experience and training of TEPCO employees has been highlighted in the investigation reports [12, 13].

4.5.5 Insufficient safety culture within TEPCO

The Fundamental Safety Principles by IAEA [11] states in its first principle that "The prime responsibility for safety must rest with the person or organization responsible for facilities and activities that give rise to radiation risks." In other words, the prime responsibility for nuclear safety is assigned to individuals and companies in the nuclear business.

And principle 3 states, "Effective leadership and management for safety must be established and sustained in organizations concerned with, and facilities and activities that give rise to, radiation risks" to emphasize the importance of stationing the safety culture within the organization.

Principle 3 lists the following three points as part of safety culture:

- Individual and collective commitment to safety on the part of the leadership, the management and personnel at all levels
- Accountability of organizations and of individuals at all levels for safety
- Measures to encourage a questioning and learning attitude and to discourage complacency with regard to safety

TEPCO's preparation against nuclear disasters that could cause serious damage to the reactor core was insufficient, and it had not made satisfactory preparation against the risk of tsunami attack over the design standards. Its capacity in responding to emergency situations was weak, education and training for severe accidents were insufficient, and their accident plan had flaws.

A portable radiation dosimeter with an alarm is indispensable for dosage management for nuclear plant workers. Fukushima-1 had about 5,000 of them; however, most of them were broken by the tsunami. As a result, not all of the workers had dosimeters, and TEPCO had many workers work without them for about a month. On March 12th to early 13th, 500 dosimeters were shipped to Fukushima-1 from Kashiwazaki-Kariwa NPP. However, TEPCO employees did not correctly identify the shipment and make use of the dosimeters.

Our investigation committee invited five experts from overseas for an international peer review on February 24 to 25, 2012. Among the five guests, Mr. Lars-Erik Holm, Board Director, Swedish National Board of Health and Welfare, highlighted this problem. Mr. Holm criticised TEPCO for (1) having contractor employees work for several weeks after the accident without dosimeters, and (2) the fact that dosimeters had been on-site.

TEPCO also had problems in terms of communicating information to the public. For example, during the press conference at 19:00 on March 11th, it repeated the wrong information that both Units 1 and 2 were under cooling and gave out only limited information about detection and leakage of radioactivity, which was of most interest to the residents. TEPCO only publically admitted that there was core damage from the accident on May 12th, two months after the accident. These problems in communication undermined confidence in TEPCO and its response to the accident.

4.6 Conclusions

TEPCO clearly failed to ensure public safety while it was running its nuclear power business. This chapter pointed out its weaknesses in preparation, which led to inadequate decisions in the accident response. Fundamentally, the company underestimated the risk of possible tsunamis that exceeded its design criteria and failed to plan sufficiently for such extreme events.

Sufficient training on accident response was lacking. The lack of training increased the likelihood of making poor decisions during the real accident when employees were under pressure. That response may have been more effective if TEPCO had conducted realistic accident drills.

Building a sufficient safety culture within an organization is not a task that can be completed in months. For corporations of the size of TEPCO, it takes years of continuous efforts on the part of top management, middle management, and employees to accomplish.

Not only those in the nuclear power business, but all entities involved with products or services that can cause threats to public safety and health must take training of their people seriously so that a real culture of safety can be built from within. This is a lesson not just for TEPCO, but applies to all of us in Japan. We all have much to learn from the 2011 Fukushima Nuclear Power Plant Accident.

References

[1] Accident Management: Measures Against Severe Accidents at Light Water Nuclear Reactor Facilities for Power Generation (see Chap 3) NSC.
[2] Accident Management Development Report, TEPCO, May, 2002.
[3] Accident Management Development Effectiveness Evaluation Report, TEPCO, May, 2002.
[4] Research report on earthquake tidal wave disaster prevention plan for Pacific coast division Ministry of Construction, the Ministry of Transportation, the Ministry of Agriculture, Forestry, and Fisheries, March 1997.
[5] Tsunami Assessment Method for Nuclear Power Plants in Japan Tsunami Evaluation Workgroup, JSCE, Feb. 2002.
[6] Act on Special Measures for Earthquake Disaster Countermeasures, 1995.
[7] Long Term Assessment of Seismic Activities from Sanriku Offshore to Boso Offshore, The Headquarters for Earthquake Research Promotion, July 2002.
[8] Regulatory Guide for Reviewing Seismic Design of Nuclear Power Reactor Facilities NSC, 2006.
[9] Numerical Simulation of the year 869 Jyogan tsunami in Ishinomaki-Sendai Plain, Kenji Satake, Yuichi Namegaya, and Shigeru Yamamoto, the 2008 Annual Report of Active Fault and Paleoearthquake Researches.
[10] Fukushima Nuclear Accidents Investigation Report, TEPCO, June 20, 2012. http://www.tepco.co.jp/en/press/corp-com/release/2012/1205638_1870.html (accessed 20.08.13).
[11] The Fundamental Safety Principles, IAEA, 2006. http://www-pub.iaea.org/MTCD/publications/PDF/Pub1273_web.pdf (accessed 20.08.13).
[12] Investigation committee on the accident at the Fukushima Nuclear Power Stations of Tokyo Electric Power Company, Interim report. http://www.cas.go.jp/jp/seisaku/icanps/eng/interim-report.html, 2011 (accessed 20.08.13).
[13] Investigation committee on the accident at the Fukushima Nuclear Power Stations of Tokyo Electric Power Company, Final report. http://www.cas.go.jp/jp/seisaku/icanps/eng/finalreport.html, 2012 (accessed 20.08.13).

Evacuation and decontamination in response to the Fukushima nuclear power plant accident

5

Y. Hatamura
Hatamura Institute for the Advancement of Technology, Tokyo, Japan

5.1 Introduction

The Fukushima accident had no direct fatalities caused by radiation released initially to the environment from the site. However, most residents, without access to accurate information, had to follow evacuation orders that kept expanding their ranges. Difficulties in evacuating the elderly in hospitals and nursing homes led to a large number of fatalities before they even reached their sanctuaries. In all, as many as 160,000 residents were suddenly required to leave their homes and over 3 years after the accident as of April 2014, hardly any of them have been able to return to their homes. Since the accident, many have passed away while living as evacuees. These are the so-called earthquake-related deaths.

Nuclear power plant (NPP) accidents cause widespread damage—not just to the plant itself, but radiation leakage to outside the plant not only damages the health of people in the neighborhood but it also damages families, communities, and even society (Figure 5.1). Our lack of proper understanding of the total picture about what would happen after an NPP accident left us unprepared in reducing the severity of damages.

Those responsible for nuclear safety were insufficiently prepared to deal with the spread of radiation in the case of a nuclear accident. Although emergency drills had been conducted, they were not anywhere near sufficient, as was found when the real emergency hit.

There is a broader responsibility of society to ensure citizens are sufficiently educated about nuclear power generation and radiation to make informed choices about government and company safety measures as well as to respond effectively to an emergency. We should include lessons about nuclear power generation and radiation in our compulsory primary educational system so the people can make proper judgments.

This chapter discusses different types of harm the residents near Fukushima-1 had to suffer (i.e., the ongoing life as evacuees, decontamination) and basic knowledge about radiation to properly understand nuclear power generation.

The 2011 Fukushima Nuclear Power Plant Accident. http://dx.doi.org/10.1016/B978-0-08-100118-9.00005-X

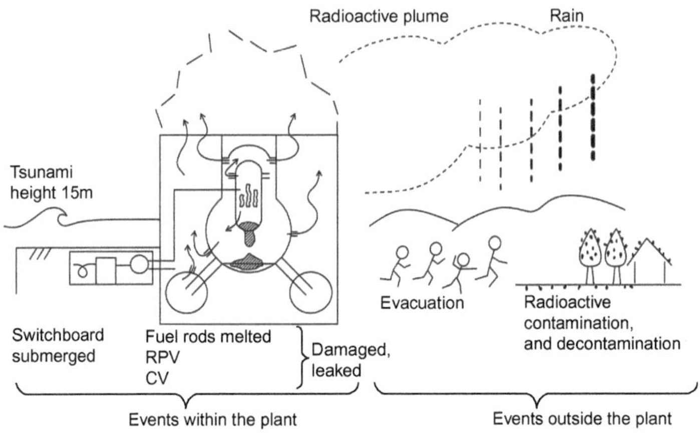

Figure 5.1 Events during the Fukushima NPP accident.

5.2 Evacuation of local residents

5.2.1 Flaws in evacuation procedure

The biggest flaw with the evacuation triggered by this accident was the lack of advance planning about what actions would be necessary with the situations in the surrounding areas. To worsen the unpreparedness, as discussed in Chapter 3, the most important spot to gather information from the accident site and make judgments, the Offsite Center, was paralyzed, and thus the evacuation proceeded without the local emergency response headquarters function.

Evacuation orders from the central government were incomplete without sufficient information, and moreover, they failed to reach all the local governments in the evacuation zones. Local regions faced hardship in trying to counteract the widespread effect of the nuclear disaster, and to add to that, the earthquake and tsunami had destroyed all means of communication. It was another problem that each local government had to take action against in isolation.

5.2.2 The process of evacuation in response to the accident

For proper understanding of the residential evacuation, we need to review the time progression of events within the power plant and the amount of radiation release to the outside to relate them to evacuation orders from the government nuclear emergency response headquarters (NERHQ). Figure 5.2 plots the relationships. The following paragraphs outline the relationship between on-site events and the progress of evacuation.

At 19:03 on March 11th, Prime Minister Kan declared a state of nuclear emergency following loss of all AC power and failure of water injection by emergency core cooling facilities at Fukushima-1 and established NERHQ. Following this declaration, Fukushima prefecture ordered evacuation of the residents within 2 km of the plant at 20:50.

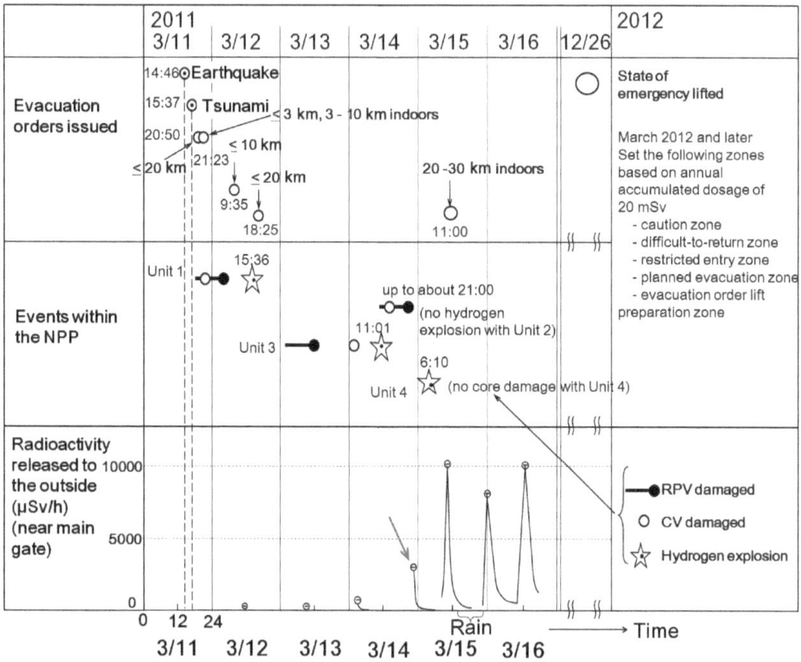

Figure 5.2 Evacuation and events in and out of the nuclear power plant.
Note: The radioactive dosage graph has an axis different from that in Figure 2.36. The vertical axis in Figure 2.36 is in log scale, and the low level of radioactivity dosage from March 12th to 14th compared to later days is not that obvious, whereas this figure clearly shows the high level of radioactive dosage from the night of March 14th onwards.

NERHQ, with the possibility of having to vent the CV to avoid core damage, took a precautionary measure and at 21:23 ordered evacuation from within 3 km of the plant and ordered people to stay indoors in the 3-10 km range. Later with the rise of CV pressure with Unit 1 and failure to vent Units 1 and 2, NERHQ decided to extend the evacuation range to within 10 km at 05:44 on March 12th.

The Unit 1 reactor building exploded at 15:36 on March 12th, and initially not knowing the cause, Prime Minister Kan took heed of the words by the Chairman of the Nuclear Safety Commission (NSC) of possible criticality, and NERHQ further extended the evacuation range to within 20 km and ordered so at 18:25.

The situation at each unit continued to worsen, and the explosions in reactor buildings of Unit 3 at 11:01 on March 14th and Unit 4 at 06:10 on March 15th led to the 11:00 order on March 15th for the residents in the 20-30 km range to stay indoors (Figure 5.3).

Local Response Headquarters at the off-site center should be in charge of ordering evacuations; however, as discussed in Chapter 3, the paralyzed off-site center forced decisions to be made at the prime minister's office. The cooling status of each reactor and other factors contributed to the decisions. However, each local region within the evacuation range was simply ordered to evacuate, and the local governments with information only available from TV and radio had to decide on the evacuation routes and guide the people in their districts.

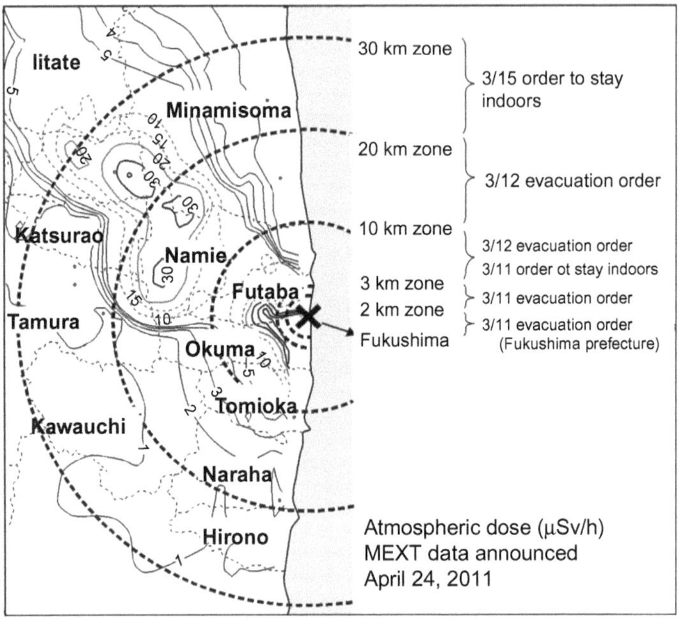

Figure 5.3 Fukushima-1 evacuation orders issued by Fukushima prefectural government and NERHQ during 3/11 to 3/15. (MEXT: Ministry of Education, Culture, Sports, Science and Technology)

Radioactivity is invisible and odorless, and people cannot sense its existence. Therefore, without adequate information, we cannot make any judgments in response to its behavior. In fact, when situations reached the point when external radiation release had turned into a threat, orders to the local governments lacked guidance on whether to stay indoors or to evacuate, and further in case of evacuation, without details of when, where, and how. At the time, even information about radioactivity spread that local governments had to rely on was unavailable.

Without accurate information, the residents were, in response to the expanding evacuation zone, merely told to move away from the plant as far as possible. It was only natural that those that had to evacuate twice and even three times felt that they were being "tossed around" by the authorities.

5.3 Predicting and monitoring the leak of radioactive material after the Fukushima accident

5.3.1 Release and spread of radioactive material during evacuation

Although Unit 2 had no hydrogen explosion, its core is estimated to have melted down (core damage), and with the damaged CV, Unit 2 probably released the greatest amount of radiation among all units. Readings from the off-site environmental radiation-monitoring posts were moderate from March 12th to 14th; however, they reached very high levels from the night of March 14th to 15th. It was the time when the core and CV of Unit 2 were damaged. The radiation around the main gate of Fukushima-1 in the morning of March 15th reached as high as 10,000 μSv/h, a level that would cause the cumulative radiation to reach 100 mSv within 10 hours. Epidemiological studies of the atomic bomb victims have shown that a radiation dose of 100 mSv or higher increases the risk of cancer.

Radiation does not necessarily spread concentrically around the origin. Its irregular spread depends on weather and terrain. In the afternoon of March 15th, when a large amount of radiation was released, the wind started blowing to the northwest, and to make things worse, rain then caused fallout with strong radioactivity in the region approximately 50 km northwest of Fukushima-1. The evacuation orders, immediately after the accident without accurate information about the spread of radioactivity, were made in sequence of larger concentric areas from the plant. Later investigation, however, revealed areas with high concentration of radioactivity formed a shoe-sole-like shape, with Fukushima-1 at the heel, covering as far as the outer rim of Iitate, a village located about 50 km away from the plant in addition to Okuma and Futaba, where the plant is located. As the authorities learned more about the radioactivity spread, they reshaped the evacuation zones to form the same shape. On April 22nd, authorities declared the area within 20 km from Fukushima-1 an "Exclusion zone," and everybody was evacuated. Outside this area to within about 50 km was partially, where the shoe-sole covered, declared a "Planned evacuation zone." Areas with minor contamination with a distance of 20-30 km from the plant were declared "Emergency evacuation preparation zone" (Figure 5.4).

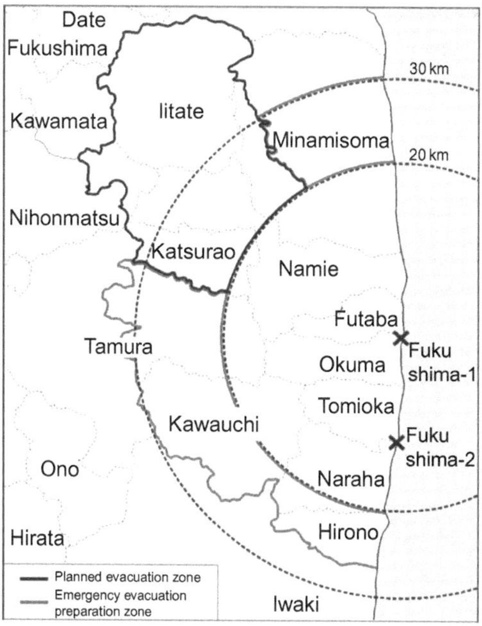

Figure 5.4 Zones for planned evacuation and emergency evacuation preparation.
Data source: April 22, 2011, NISA announcement.

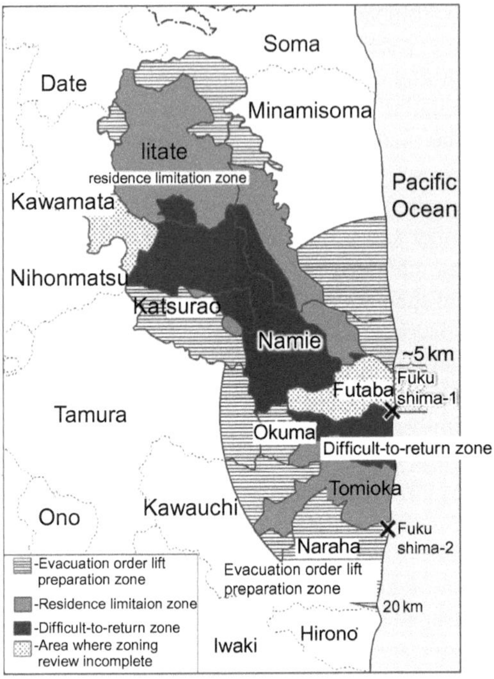

Figure 5.5 Evacuation order zone and caution zone as of March 8, 2013.

About 9 months after the accident on December 16, 2011, the authorities judged the reactor to have stabilized reaching the state of cold shutdown, and on December 26th, with the threshold of the annual accumulated dose of 20 mSv, they divided the area into a "difficult-to-return zone," "residence limitation zone," and "evacuation order lift preparation zone." The zoning was reset on March 7, 2012 after review (Figure 5.5).

5.3.2 *Speed of radiation spread*

Let us explain the terms and units we have been using so far in this book about the strength and magnitude of radiation.

Sievert (Sv) is the unit for the dose of radiation that affects the human body. The unit milliSievert (mSv) that we see more often is its 1 thousandth, and microSievert (μSv) its 1 millionth. For the impact of radiation on human health, what counts is the total amount of radiation (cumulative dose) the body is exposed to. The intensity of radiation, "air dose," is expressed with the amount of radiation dose per each hour while the body is at the location. We use microSievert per hour (μSv/h) for hourly exposure and milliSievert per year (mSv/y) for annual exposure. Cumulative dose and air dose are related; e.g., when exposed to 1 μSv/h for 1 year, the total cumulative dose for the year is: 1 μSv/h × 24 hours × 365 days ∼ 9 mSv.

As we showed earlier, the area of high air dose is spread in the shape of a shoe-sole, and some spots like Namie and Iitate, that are 30 km away from Fukushima-1 plant, have an air dose as high as 10 μSv/h or higher (as of April 2011). An air dose of 10 μSv/h will result in an annual cumulative dose of 87.6 mSv, and if a cumulative dose that exceeds 100 mSv causes damage to the human health, this 87.6 mSv is a fairly high value.

Ministry of Education, Culture, Sports, Science and Technology (MEXT) published a "Distribution map of radiation dose around Fukushima Dai-ichi & Dai-Ni NPP (as of January 11, 2012)" [1] that estimated the total radiation exposure from March 11, 2011, to March 11, 2012. The document shows a 20 mSv area extending as far as 50 km from Fukushima NPP. Although very unlikely, a person who stayed in the area for 5 years would probably suffer effects of radiation.

Radiation released from NPP is carried by wind (called radioactive plume), and if it rains, then the raindrops will capture the radioactive material and take them down on the ground. In most cases, however, without rain, wind will carry radiation to distant locations and disperse it.

5.3.3 System for Prediction of Environmental Emergency Dose Information (SPEEDI) simulation of radioactive spread

Nuclear Safety Technology Center (NUSTEC), an affiliate company of MEXT, started operating a System for Prediction of Environmental Emergency Dose Information (SPEEDI) in 1986 for use in evaluating methods to protect people from radiation effects.

SPEEDI finds "when," "in which direction," and "in what quantity" radiation released from an NPP will spread based on information from the radiation source, terrain, and weather. The system works with radiation release data transferred from TEPCO via NISA to NUSTEC for the simulation, and the results are forwarded to NISA, affected prefectures, the off-site center, and NERHQ.

SPEEDI simulated the reaction to the 1999 JCO criticality accident, and the same task was to be performed for this Fukushima-1 accident. This time, however, the earthquake and tsunami caused damage that disrupted the transfer of release source data to carry out the simulation.

Figure 5.6 shows the SPEEDI calculation from assumed release source data on March 15, 2011, which was the day with the largest amount of radiation release. The air dose map (Figure 5.7) from aircraft monitoring, made available 2 months after the accident, shows a shoe-sole-like area of high dose to back the simulation results by SPEEDI.

In the case of this accident, however, even though the direction of radiation spread had been calculated, the residents had to evacuate without knowing how much radiation was spread in which direction just because the radiation release source data were unavailable.

5.4 The impact of nuclear radiation on the human body

5.4.1 Risk factors of nuclear radiation

People live with a number of factors that can harm health (Figure 5.8a). Risk factors that can harm people include cigarettes, alcohol, other lifestyle habits, and social stress. Radiation is another one of such factors. The concern with this Fukushima

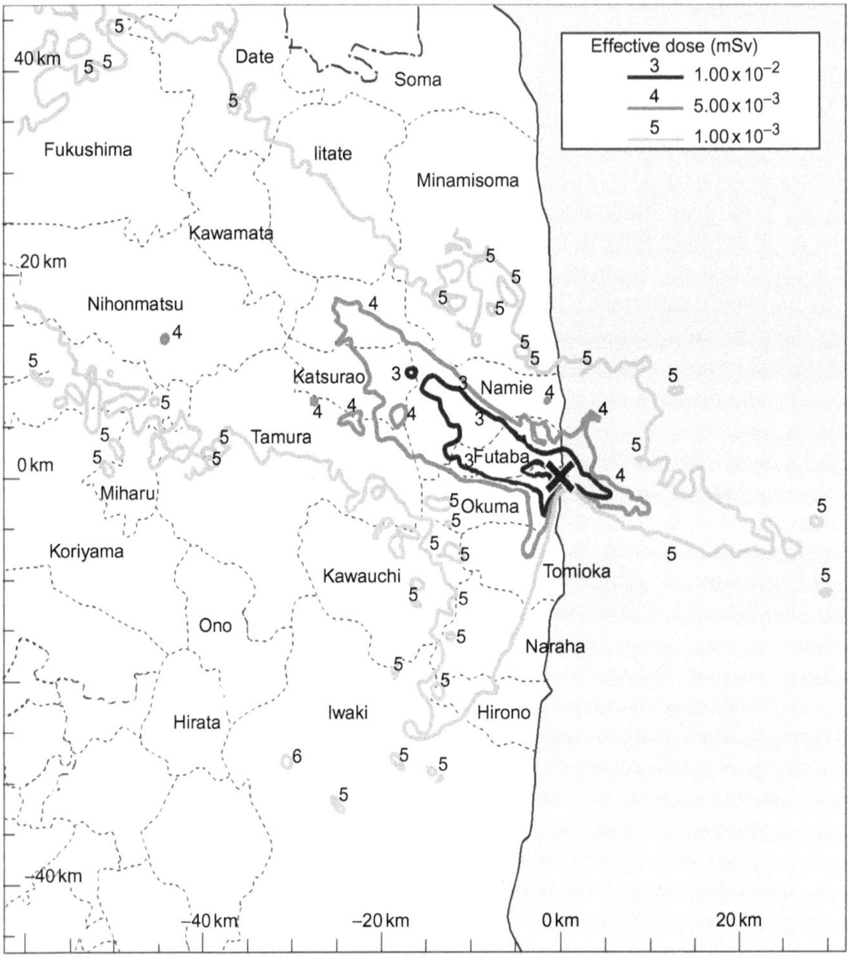

Figure 5.6 Simulation results from SPEEDI by NUSTEC office of MEXT. Effective dose from external exposure during March 15th, 09:00 to 16th 09:00.

accident is the effect of radiation on human health. Here we will explain the basic knowledge of the relationship of radiation to human health and then discuss the impact of radiation from this accident.

We have to start by classifying radiation that affects human health in two groups. One is "natural radiation" from radioactive material that existed in the world from the beginning. The other is "artificial radiation" from radioactive material placed in the environment by accidents, atomic bombs, and other man-made causes.

Natural radiation exists in cosmic rays from outer space, earth (rock beds and stones), and some other natural objects. This causes external exposure from outside the body and internal exposure through breathing, eating and drinking. Humans are always exposed to natural radiation and cannot hide from its influence.

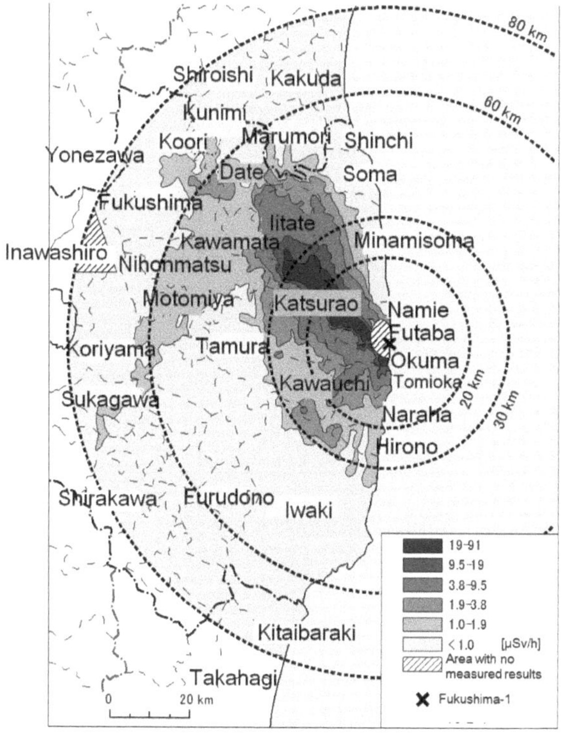

Figure 5.7 Results from MEXT and aircraft monitoring by U.S. Department of Energy. Atmospheric dose at 1 m above ground, converted to values on April 29, 2011.

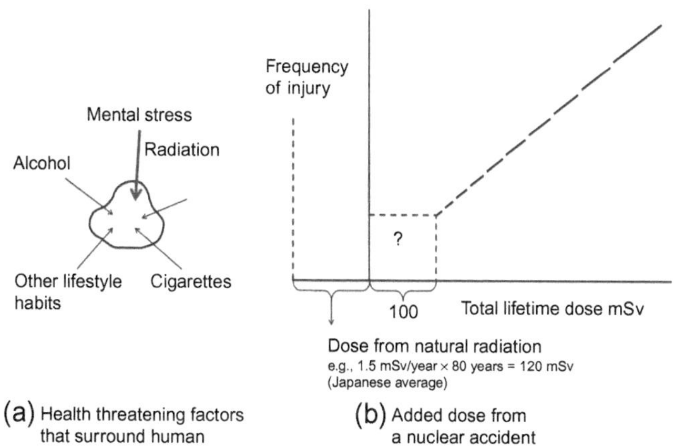

Figure 5.8 Effect of radiation on human health.
High Energy Accelerator Research Organization, Radiation Science Center, March 2005.

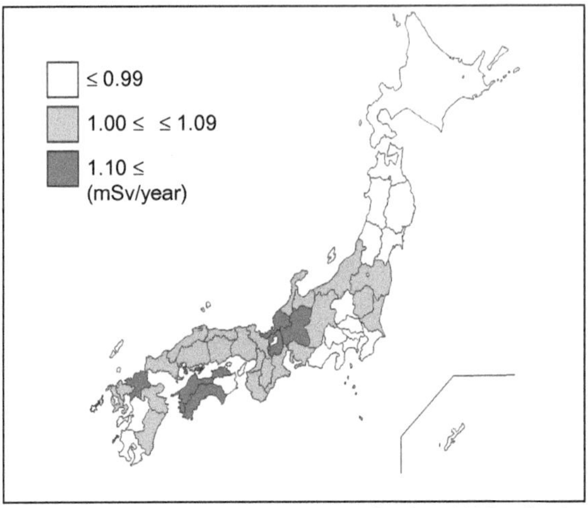

Figure 5.9 Annual dose from natural radiation in Japan.
Source: High Energy Accelerator Research Organization, Radiation Science Center, March 2005.

The amount of natural radiation differs geographically, e.g., an average annual dose from natural radiation in Japan is about 1.5 mSv with the amount in western Japan being about 1.5 times that in eastern Japan (Figure 5.9). The annual world average is slightly higher at 2.4 mSv.

If one lives for 80 years in Japan, the accumulated dose is 1.5 mSv/y × 80 years = 120 mSv. The effect of this natural radiation on the health is not clear; however, there is no noticeable regional difference within Japan (we will discuss external and internal exposure later).

The concern with this accident is the effect of exposure from the radioactive material it released. We have to add the amount due to the accident to the natural radiation in the area.

5.4.2 The impact of nuclear radiation on human health

In evaluating the effect of radiation from the accident, we have two methods: one is to find the dose rate with μSv/y, mSv/y, μSv/h, or mSv/h, and the other is to count the total amount of radiation on one person (accumulated dose). The total dose with a short exposure to high dose and long exposure to low dose are both called accumulated dose. Even if the two cases had the same accumulated dose count, the impact on human health is totally different with instant exposure and an exposure over the whole life.

None of the on-site workers or off-site residents suffered acute radiation syndrome from a large dose within a short period of time. The concern, thus, is long-term effect of chronic dose (i.e., a small amount of radiation received over a long period of time).

For humans, we know that an accumulated dose of 100 mSv or more of extended chronic dose increases the risk of cancer (Figure 5.8b).

If the accumulated dose is 100 mSv or less, however, the impact of radiation dose on health is unclear. The reason is that radiation is then just another one of all the risk factors that surround us, and we cannot identify the pure effect of radiation alone.

One may think that as long as the impact of 100 mSv or less accumulated dose on health cannot be denied, we should keep the exposure as low as possible. That is a good practice if circumstances allow it; however, an overly nervous reaction against exposure can cause mental stress and add other risk factors that may affect human health even more.

We have to be aware that we are constantly exposed to natural radiation, and chronic dose is just another risk factor that may affect human health.

5.4.3 Physical and mental effects from nuclear radiation

There are two ways for humans to be exposed to radiation. One is external exposure (i.e., the effect from radioactive material to the outside of the body). The other is internal exposure from radioactive material taken inside the body through breathing or food and beverage consumption. Both external and internal exposure cause physical effects, mainly in the form of damage to the DNA or cancer.

We should not forget, on the other hand, that the mind is also affected by radiation, in addition to physical damage. The most common form of mental damage is the fear of risk to the body (e.g., of developing cancer).

This fear causes people to overreact to radioactive material by making irrational decisions and acting unreasonably. When the whole society is influenced, it gives rise to harmful rumors and can hamper proper evacuation and decontamination. To minimize the impact on the human mind, we need to have an understanding of how radiation affects human health.

Each and every resident from the nuclear accident affected area, in order to lead a life with a calm mind, needs to have their health monitored with proper treatment as needed. Long-term health monitoring is necessary. The collected data will serve as a valuable source of information in understanding the relationship between humans and radiation.

5.4.4 Effects of internal exposure to nuclear radiation

Radioactive material taken into the human body through breathing, consuming food or beverages, or through skin, decays inside the body and emits radiation that affects our cells and DNA. This phenomenon is called internal exposure. The amount of radioactive material taken inside the body reduces over time by being discharged from the body.

This accident released a large amount of radioactive iodine to the outside. Radioactive iodine easily evaporates to spread over a wide volume in the atmosphere and quickly invades the body through breathing, eating, or drinking. Iodine taken inside the body accumulates at the thyroid gland to cause internal exposure. This organ is in the process of growing in infants and is highly active in young people. Children can easily suffer from a thyroid disorder when radioactive iodine accumulates in the thyroid gland.

Filling up the thyroid gland with stable iodine will prevent further accumulation of iodine there, and excessive iodine that arrives later is quickly discharged with urine. This is why taking stable iodine tablets is effective in preventing accumulation of radioactive iodine in the thyroid gland. Iodine, when it enters the body, quickly reaches the thyroid gland; thus, it is important to take the tablets before exposure. It is also said that a diet that includes plenty of seaweed will increase the amount of iodine in the body and thus will block radioactive iodine from reaching the thyroid gland.

Much has been said about iodine exposure with this accident; however, the amount of release and its spread have not really been clarified. The short half-life of 8 days with iodine 131 makes it difficult to take any measurement after time has elapsed after the accident. We can probably only make estimates of the iodine release and distribution based on measurements made immediately after the accident and recorded data of other radioactive material with longer half-lives.

Cesium, on the other hand, dissolves in water; thus, when it enters the body, it reaches and stays in the muscle tissues to cause internal exposure. Its biological half-life (time it takes to be discharged from the body until the radioactivity is half the original magnitude) is about 70 days. For cesium, we currently do not have the equivalent of stable iodine tablets, so we have to take care not to let cesium into the body with food. This is one of the contributing factors that caused the harmful rumor about Fukushima produce.

5.4.5 Effects of external exposure to nuclear radiation

Humans have always been exposed to natural radiation from cosmic rays that reach the earth from outer space and radiation from radioactive materials in the natural environment.

Regions where radioactive material reached were subject to added radiation from the accident to the natural radiation of 1.5 mSv/y. If the accident had added 20 mSv/y, the total amount would have reached 21.5 mSv/y, and the people would have been exposed to 14 times the natural radiation. On the other hand, if the addition caused by the accident was 1 mSv/y, adding it to the original 1.5 mSv/y would only amount to 2.5 mSv/y. The total amount of 2.5 mSv/y is about the same as the world average of annual natural radiation. Decontamination efforts to lower the added dose to about 1 mSv/y are reasonable.

When the authorities discussed the criteria for using school facilities, the value 3.8 μSv/h had a strong message. An outdoor exposure of 3.8 μSv/h with 8 hours of outdoor activity with indoor exposure for the rest of the day, where the exposure is about 0.4 times that from outdoors, sum to a daily exposure of 54.72 μSv. This value multiplies to an annual exposure of 20 mSv. This is how the criterion of 3.8 μSv/h was reached for judging whether schoolyards were acceptable for activities or not.

When we look at external exposure, we must turn our attention, among the radioactive material released by the accident, to cesium 137 with a long half-life of 30 years.

Iodine 131 and cesium 134 have relatively short half-lives of 8 days and 2 years, respectively, and their effects do not last very long. Cesium 137, however, with its long half-life of 30 years, once released to the environment, causes long-term effects that render land and houses uninhabitable. This is why cesium 137 is of concern regarding decontamination.

Radiation, however, is just one factor among all others that affect human health. Being overly concerned with the hard-to-accomplish limit of 1 mSv/y and forgetting about other factors will cause us to make the wrong judgment.

5.5 Assessing the evacuation process in response to the Fukushima nuclear power plant accident

5.5.1 Stable iodine tablet distribution and its administration

The main ingredient of stable iodine tablets is nonradioactive iodine, which prevents buildup of radioactive iodine in the thyroid gland if taken before internal exposure to prevent its damage.

Administering stable iodine tablets is carried out according to procedures set out by the Nuclear Safety Commission of the government advising the local NERHQ, which makes decisions for prophylaxis administration and reports it to central NERHQ. The decision by the headquarters is delivered to each prefecture governor, who then relays the decision to the residents.

The six counties around Fukushima-1 and -2 had stocks of stable iodine tablets following the "Emergency Radiation Exposure Medical Action Procedure" [2]. The prefecture, on March 14th, decided to distribute stable iodine tablets to all residents under age 40 living within about 50 km of the plant and completed the distribution by the 20th.

The largest radiation release, however, took place on March 15th, and prophylaxis distribution after when it was needed most had no point. This event just teaches us how inappropriate the procedure was.

On the 16th, the local response headquarters ordered "stable iodine tablet administration when leaving the 20 km zone" to the prefecture and 12 related counties; however, the prefecture did not make orders for prophylaxis distribution because the 20 km zone had already been evacuated.

In addition, on March 13th, the Nuclear Safety Commission advised administering stable iodine tablets to those with exposure doses over a limit during the screening. The advice, however, did not make it to the local response headquarters.

Some counties around Fukushima-1, however, made their own decisions to distribute stable iodine tablets to their residents. As we discussed in Chapter 3, Miharu, based on the rise of radiation at Tohoku-EPCO Onagawa NPP on March 14th and the weather forecast on March 15th, decided to distribute and administer stable iodine tablets.

Officers of Fukushima prefecture, who learned about the distribution and administration, ordered Miharu to stop the distribution and recover those already given out for the reason that the central government had not given orders to do so. The county did not follow the halt and recall order.

Such confusion was the result, as we will discuss later, of operations run by people with poor understanding of stable iodine tablets and their role in radiation protection.

5.5.2 Inadequate evacuation process

When a large amount of radiation was released to the outside on March 15th at 11:00, authorities ordered residents in the 20- to 30-km range from the plant to take shelter indoors. The city of Minamisoma led an evacuation of the residents that elected to leave the city. There were three routes that extended out from the city. The southbound route passed near Fukushima-1, and the northbound route ran along the coast, which might have been blocked by damage from the earthquake and tsunami. Most of those who opted for evacuation, thus, took national route 114 (a.k.a. Tomioka Ocean Road) that extended northeast from Fukushima-1 toward Iitate and Kawamata (Figure 5.10).

At the time, however, the wind was blowing to the northeast, and the rain worsened the situation by dropping radiation in the 20- to 30-km section northeast of Fukushima-1 at levels higher than other areas. The fleeing residents, without knowing the situation, ran toward the area with highest level of radiation fallout.

The radiation spread would have been known if the SPEEDI simulations were made available. The residents could have made proper judgments if the direction of escape and the time were right. Without this crucial information, the residents of the city of Minamisoma, who were only informed that danger was on its way, decided to flee the area instead of hiding indoors. If they had stayed indoors at this

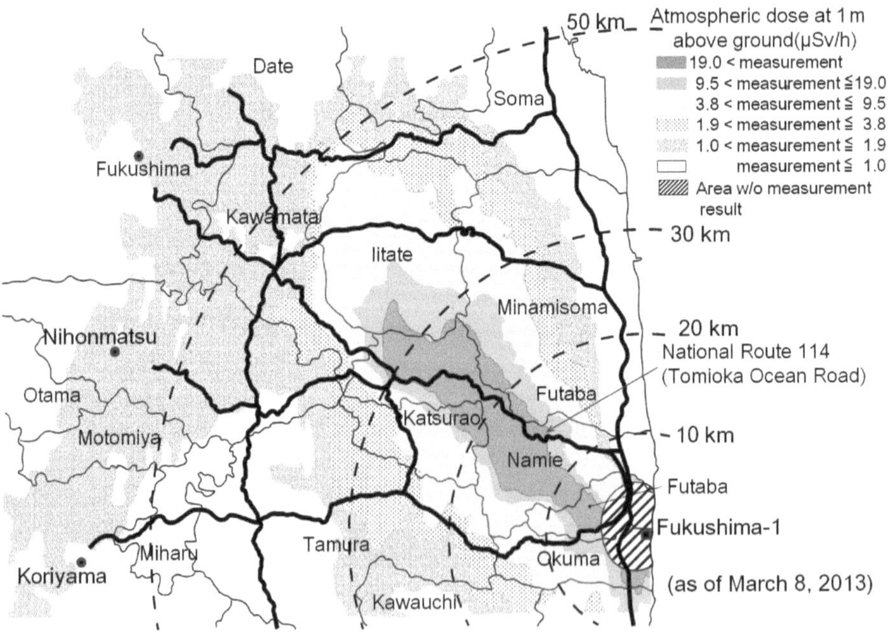

Figure 5.10 Evacuation route and atmospheric dose. Atmospheric dose from aircraft monitoring on April 29, 2011, for MEXT distribution map of radiation dose.

time, effects of radiation would have been smaller, and even if they had decided to leave anyway, they could have gone in the right direction with information from SPEEDI. The residents learned the facts later and were furious about what happened. Their anger comes as no surprise.

5.5.3 Evacuation routes within each county

Without prior evacuation plans to counter nuclear disasters, evacuation orders were issued and the counties had to make their own judgments and act accordingly.

After the accident, Futaba, immediately north of the plant, and Okuma, at the south, were evacuated. The evacuation order stated to "get out" without telling the destination. Some regions were lucky to use buses arranged by the central government, but most counties prepared means of transportation on their own and fled away from the plant.

As the evacuation zone extended, many evacuees had to make two or more relocations. Futaba had to move from one place to the next, and many of its residents ended up in the far city of Kazo in Saitama prefecture. People of Okuma also passed through multiple counties to reach Aizuwakamatsu far in the west.

Namie, with most of its population in the exclusion zone, fled to Nihonmatsu, and Tomioka ran to Koriyama. Some residents of Minamisoma near the plant travelled through regions of high radiation as we explained earlier.

Figure 5.11 shows the original residence of the evacuees and where they ended up. The figure shows the chaotic relation between the original place of residence and where the people found sanctuary. Some evacuees, in search of safe places, had to make long relocations. The figure is evidence that reveals the lack of effective evacuation plans prior to the accident.

Figure 5.11 Evacuation destination for each administrative district.

5.6 Radioactivity, radioactive material, and radiation

In Japan, we are often confused with the terms *radioactivity*, *radioactive material*, and *radiation*. Their meanings are different, and we shall distinguish their uses. Radioactivity is the capability of an atom at its nucleus to separate and generate radiation or radioactive rays in the process. Radioactive material is material with radioactivity.

For Japanese people, when we hear *radioactivity*, we think about the devastation from atomic bombs. Not just the people living near them but the entire nation lacked knowledge about nuclear power plants and radioactivity. The outbreak of this accident forced us to suddenly face them squarely.

5.6.1 Common misconceptions on radioactivity

A severe accident in an NPP means possible leakage of radioactive material to the outside. People, without proper knowledge of its impact on the human body and the environment, seem to have overreacted to all sorts of information. The overreaction made the situation worse.

Without proper knowledge, many of us make judgments and take actions based on wrong assumptions. These include harmful rumors.

Harmful rumors started with fresh food products. Many believed produce and seafood from the Fukushima and Tohoku areas were contaminated with radioactive material and avoided them even if their safety was confirmed by inspection. People thought that was the way to protect their own health.

Another form of harmful rumors was other parts of the country refusing to dispose of rubble that the tsunami left behind. Some even thought radioactivity transfers to others through physical contact with the evacuees, repeating the same rumors that broke out at the time of the 1999 JCO accident.

To prevent such harmful rumors caused by wrong knowledge, we have to arm ourselves with the right knowledge about radioactive materials, especially those that are released by nuclear reactor accidents.

5.6.2 Radioactive half-life

To learn about radioactive material, we need to understand "half-life." Radioactive material decays over time and changes into material that has no radioactivity. Half-life is the time it takes a specific radioactive material to decay (the phenomenon of an atom emitting radiation to turn itself into another type of atom) until the amount of its radiation is half the original size. Half-life is different with each radioactive material, e.g., Plutonium, with a large atomic weight, hardly dissolves in any liquid, and it has a long half-life of about 20,000 years; it almost emits radiation permanently.

There were three radioactive materials of concern that spread to the outside with this accident; iodine 131 with a half-life of about 8 days, cesium 134 with about 2 years, and cesium 137 with 30 years. We will later discuss iodine 131 and cesium 137 because their amounts, behaviors, and sizes of effects were of concern.

5.7 Benefits and challenges of evacuation

In areas where people have been forced to evacuate due to an accident, they have to face disruption to their lives, families, workplaces, local societies, and their well-being.

The issue that causes the most distress to humans is a huge change in the environment that surrounds them. The evacuees suffered a greater impact from the catastrophic changes in their environment than from the radioactive material. The evacuees have been subject to continued physical and mental distress from the experience of being evacuated. Such continuous extended evacuations are the biggest negative influence on people's well-being.

This fact is clearly seen with the "earthquake-related deaths." As Chapter 1 showed, the number of earthquake-related deaths from immediately after the disaster to the end of September 2012 was slightly over 2300 nationally, and almost half of them, 1121, were deaths in Fukushima prefecture. Some were attributed to the suffering from the earthquake and tsunami; however, most of these deaths are likely to have been caused by forced evacuation or a dramatic change in lifestyle.

The impact of the nuclear accident and the following catastrophic change in environment cannot be assessed just by the fatalities. We should not forget that many of the 160,000 evacuees from Fukushima-prefecture have been burdened with health and mental problems that cannot be shown in numbers.

The 1986 Chernobyl nuclear accident also taught us about the impact of evacuation itself on humans. The then Soviet Union government forced evacuation, and the average lifespan of the evacuees was 7 years shorter than for those who stayed in the same area (Russian government report [3,4]).

Those that stayed may have received radiation effects; however, in the end, those effects were less than the physical and mental suffering caused by evacuation. This example warns us that we have to think about the overall health of people in judging their long-term evacuation, otherwise we may end up causing them more suffering.

The best solution, of course, is to have them recover their original lives; however, if that is impossible for the years to come, we have to set aside any plans of returning them home. The final judgment should be made by the evacuees themselves. For the best health and lives for them, we may want to prepare for the option of starting new lives in a new place.

5.8 Effectiveness of methods to decontaminate the site and local area

5.8.1 *Radioactive material cannot be erased*

Radioactive materials released from the Fukushima nuclear reactor floated in the air as a radioactive plume, like an invisible cloud. The wind carried the radioactive materials far and wide. As time passed, they started to fall on the ground and on leaves.

If rain falls from above the radioactive plume, clusters of radioactive material and lumps of water (raindrops) collide, and raindrops that have captured radioactive

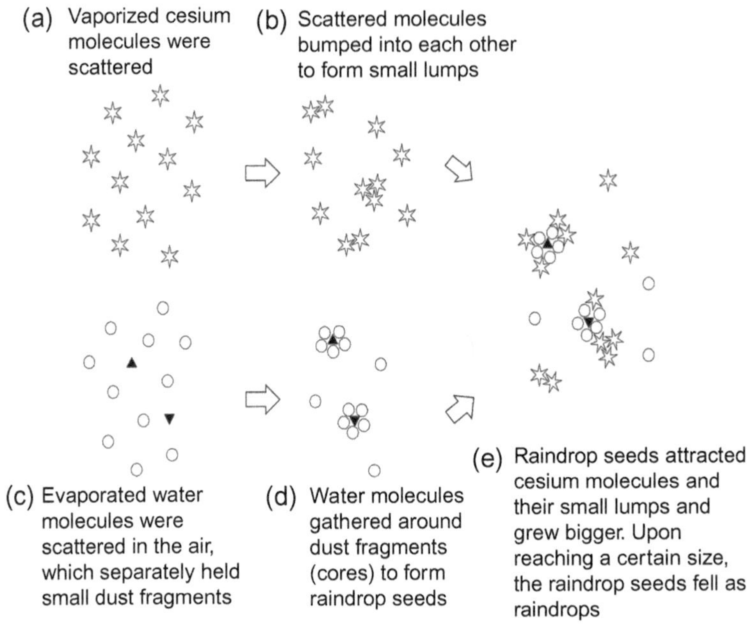

Figure 5.12 Mechanism of vaporized cesium falling as raindrops.

material fall on the earth (i.e., soil grains and surfaces of leaves). Areas highly contaminated by rain with radioactive material are called "hot spots." This is shown in Figure 5.12.

Figure 5.13 shows a sketch of what the residents of the village of Iitate, one of the hot spots, felt and thought about. The sketch is based on comments from them. The comments were "Invisible clouds of radioactivity came from the other side of the southeast hills and the 'radioactivity' fell on and got stuck to our rice patties, produce fields, houses, and forests."

Some of the fallen radioactive materials were washed away with the rainwater. That is why higher radiation was measured in rain gutters, water ditches, and naturally formed rain channels. Radioactive material that was not washed away, on the other hand, stuck to soil particles, leaves, and roof shingles and stayed there even after the water evaporated. Radioactive material molecules that got stuck to other object surfaces cannot be washed away even with brushes. Figure 5.14 shows the mechanism.

It is a natural desire to remove the radioactive material that spread out from the nuclear plant. Unlike chemically disinfecting poison with neutralizers, we do not have processes for erasing radioactivity.

The accident released several types of radioactive material, and the one that matters the most in terms of decontamination is cesium 137 with a half-life of 30 years. Cesium 131, that can accumulate in the thyroid gland, has a half-life of 8 days, and it quickly changes into a non-radioactive material called xenon. Cesium 134 has a half-life of 2 years, and so 2 years after the accident, about half of it had changed

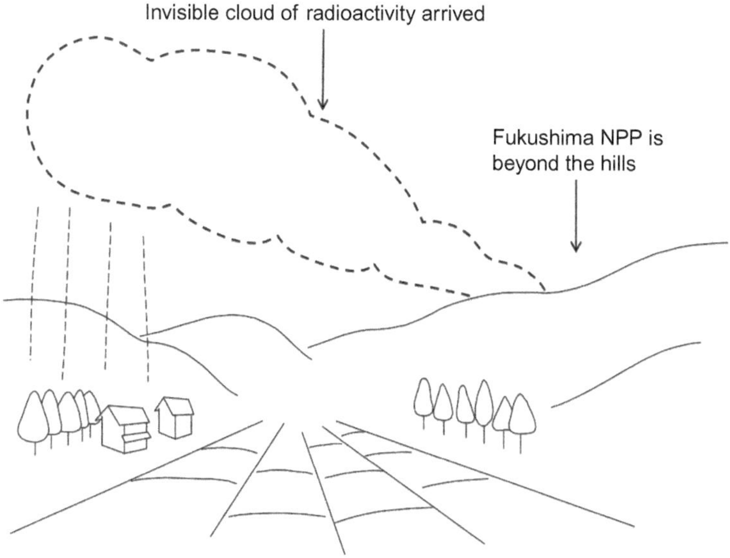

Figure 5.13 Cloud of radioactivity that reached Hiso area of Iitate-mura (sketch by Hatamura based on comments from local residents).

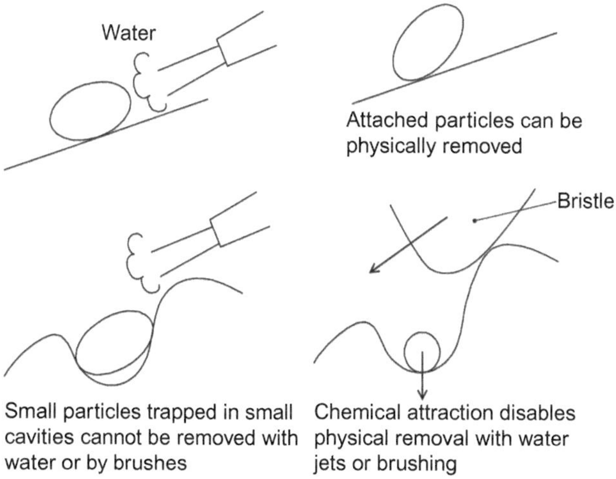

Figure 5.14 Small particles with the size of atoms cannot be removed by physical means.

into barium 134. Cesium 137 emits radiation to turn into nonradioactive barium 137. The amount of radiation with cesium 137 reduces to half the original in 30 years, a quarter in 60 years, an eighth in 90 years, and about one-tenth in 100 years. We can only count on decay of radioactivity in dealing with released radioactive material.

Under these conditions, the best practice is to acknowledge the existence of radioactive materials, take measures to minimize their impact on humans, and wait for time to pass so the radioactivity decays.

The central and local government agencies are currently working on plans to gather the soil and leaves contaminated with radioactive material and store them at designated places. The motivation is to ease people's anxiety in living close to radioactive material without thinking about the amount of radiation. If we, however, open our eyes to reality, such measures are likely to fail. This is because areas that are named for storage and others in the route of transportation are refusing to take those roles.

5.8.2 Local storage of radioactive material

Are we out of measures to handle the situation? No, not really. We are left with one method. That is to store the radioactive material "on-the-spot" so it does not affect our daily lives.

Here we will explain the actual ways of "on-the-spot disposal."

The simplest and most effect method is to build mounds of radioactive material (Figure 5.15a). For example, if we scrape 5 cm of soil off the surface of a 25 m × 25 m area and build a mound, its height will be 1.25 m (Figure 5.15a).

These mounds copy what our ancestors did in handling volcanic ash. Japan is a land of volcanoes and has a long history of dealing with volcanic eruptions. When volcanic ash hampered the farm work, people gathered the ash in one place and built ash mounds [5] and resumed their farm work. We can apply this ancient wisdom to decontamination. We can gather the contaminated soil and store it where it was scraped and wait for the radiation to decay. This method, however, leaves the contamination in sight of people in the area and can weigh on their minds.

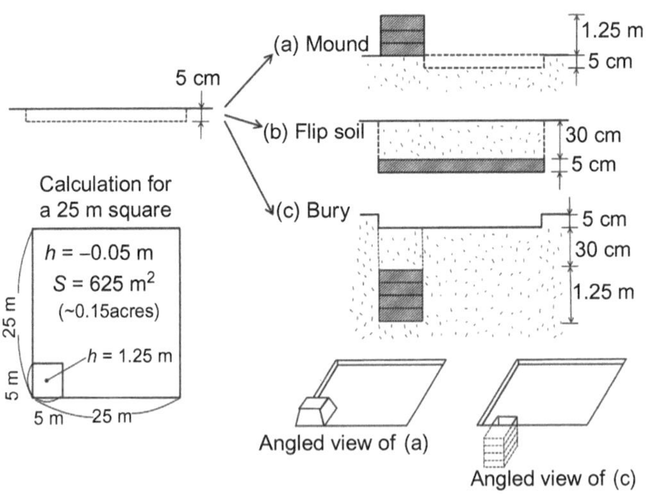

Figure 5.15 Practical handling of contaminated soil is by mounding or burying.

The on-the-spot disposal actually done in schoolyards and some other places is "soil flipping" (Figure 5.15b). This method is effective; however, the cost and labor involved are too great, and it has not been done elsewhere.

A method similar to soil flipping is "ground plowing." We can call this method "agricultural soil flipping," which is not as complete as soil flipping. Ground plowing uses a plowing machine to dig into the soil and mix the surface with the soil underneath. The method is similar to flipping. However, it can only lower the air dose by mixing the contaminated surface soil with that underneath, and it is incomplete compared to flipping, which puts the radioactive material underground.

A more effective disposal method is to dig a deep hole to bury the contaminated material and cover the surface with clean soil (Figure 5.15c).

The burying method is first to open a deep hole in one corner of the land, and put the clean soil to the side. The contaminated soil is then scraped off the surface to a depth of 5 cm and placed in the hole. Once all the contaminated soil is buried in the hole, the clean soil that was placed to the side is then replaced on the top. This method hides the contaminated material from the human living space and not only eliminates the radiation effect on humans but also removes the mental burden of having to look at it.

We now look into the practicality of this burying method in detail with the authors' assumptions. Cesium 137 (cesium in the remainder of this section) released from the nuclear reactor dissolved into water (rain), fell on the ground, and passed over the ground with water. Cesium dissolved in water travels with the water running through grains of soil. When cesium contacts the surface of soil grains, e.g., silica (SiO_2), it adheres to the crystal surface with electric force (ionic force) to make atomic bonding. Once this bond forms, the cesium does not easily dissolve into water again (Figure 5.16). In other words, the soil grains capture the cesium and do not let go.

This mechanism has been the subject of research [6]. Some evidence that supports the idea has been found in the contaminated area. Here are some examples:

- Radioactive dose is higher on the roof, in the gutter, and in concrete flow channels, and its level does not change over time.
- Brushing alone does not lower the dose level on shingles and road surfaces, unless the surface is scraped off.

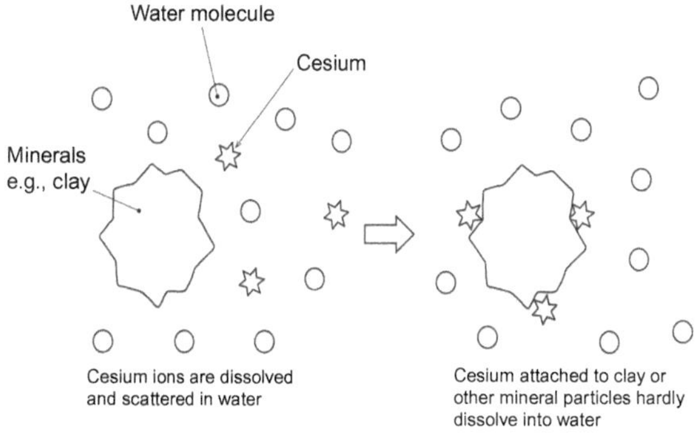

Figure 5.16 Mechanism of capturing radioactive cesium in soil.

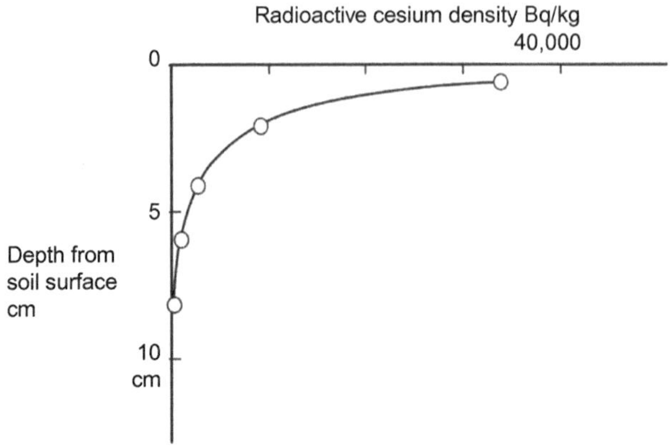

Figure 5.17 Sample vertical distribution of radioactive cesium in soil.
Based on March 14, 2012, data from Soil Science Symposium, by the Soil Science Committee of the Science Council of Japan (SCJ).

- Radioactive dose is highest at the surface and drops exponentially in the depth direction. Most of the cesium is bonded to the soil within a 3 cm depth, and thus, scraping off 5 cm of soil from the surface will remove the contaminated soil (Figure 5.17).
- The distribution remains unchanged over time and only reduces exponentially with time.
- 2 years after the nuclear accident, cesium is barely detected in water streams. Occasionally heavy rain will result in a small amount of cesium. This is because cesium is attached to small grains of soil and is not dissolved in the water.

These facts (with assumptions), observed 2 years after the accident, suggest that cesium, no longer floating in midair or water, has attached itself to solid objects like soil grains or leaves and is unlikely to dissolve out into water.

We do not have to put the contaminated materials into plastic bags if cesium will not dissolve into water and we can just bury them in deep holes so they will not affect our lives. All we have to do then is to wait for the radioactivity to decay over time.

Our ideas here have not been verified; however, we believe they are on the right track. When we look at the serious suffering in areas devastated by nuclear accidents, we need to seriously evaluate such measures. We strongly hope that practical solutions like this one will make their way to becoming counter measures.

5.8.3 Looking for a practical solution

The decontamination process now gathers all the contaminated materials in one spot with plans to move them to a remote location away from our daily lives for storage. Transportation and storage will lead to conflicts among regions, and we will not reach any conclusion. If we stick to the plan of gathering and storing contaminated material in large quantities, the decontamination will stall. Then without ways to recover their former lives, local regions will stay devastated, and their recovery will be even more difficult.

We believe, with the assumption that we cannot erase radioactive materials, the most practical solution is the on-the-spot disposal method of "burying" them. We hope that the forced evacuees will initiate their own action without waiting for decisions by central and local governments. Decontamination of the mountain forests, that cover two-thirds of the entire Fukushima prefecture, has not even started. We have to decontaminate a wide range of mountain forests so people can use their houses, schools, and office buildings with confidence. The current decontamination just of areas close to houses, schools, and office buildings is not at all sufficient.

All of us must have the right understanding about dealing with radioactive material and share the knowledge. We then have to quickly find and carry out the actions that everyone agrees to, whether it is decontamination or returning home. Otherwise, delayed decontamination will further delay the returning home and other regional activities down the road.

5.9 Conclusions: key lessons learned in the evacuation, return, and decontamination at Fukushima's nuclear power plant

In planning evacuation, return, and decontamination, we have to base our plans on minimizing damage to residents. Plans made immediately after the accident will change over time, triggered by events that take place as time passes. We should not necessarily stick to plans that we made immediately after the accident.

Damage to the local residents is not just radioactive fallout from the nuclear plant and resulting contamination of the ground. The more significant damage is the mental stress from collapse of the family, local society, loss of work, and not being able to use the contaminated ground. What we need to do now is to minimize the total damage to these people.

Having excessive fear of "radioactivity" and staying away to avoid living in the area with slightly higher dose or insisting on unrealistic decontamination plans could magnify the overall damage to people.

It is a pity, but we have to accept the fact that soil, once contaminated with radioactive material, can no longer be put completely back to the original state no matter what we do. It is impossible to eliminate radioactive material, and thus, we have no countermeasure that works on everything at once. We need to understand that we just have to follow nature and wait for the radioactivity to decay.

Having cesium 137, with a half-life of 30 long years, in the picture, the future of the affected area will be: 3 years for the return of people, 30 years for local activities, and 100 years for a complete revival.

Setting the target for the return of people at 3 years has reasons. If it takes as long as 10 or 20 years, the evacuees will build their lives in their new homes and will no longer be able to return to their homeland. People take their evacuated lives as temporary ones in terms of life, study, and work. Unless we set a time limit, evacuees will not be able to build their own life plans. That is why a period of 3 years makes a reasonable target for returning.

The period of 30 years for the revival of local activities starts from people returning within 3 years to restart their lives. After that, people will gradually rebuild their lives, and then after about 30 years, the local society will resume its activities. Other regions that suffered different types of large-scale disasters in the past took about 30 years to resume activities. If 20 mSv/y is the criteria for living, the radioactivity dose will reduce to half in about 30 years and thus, uninhabitable areas will reduce to where it currently measures 40 mSv/y or more.

The whole community of Fukushima prefecture will take about 100 years for a full recovery. As we discussed earlier, radiation from cesium 137 will drop to about one-tenth in 100 years. Areas in which the accident-induced radioactive dose is 1 mSv/y or higher will reduce, and only a fraction will remain uninhabitable. In fact, however, an additional drop in radioactive dose is expected from weather effects that will wash away some of the radioactivity. These facts led to the expectation of 100 years for the complete revival of the prefecture.

With all the preceding discussions, the environment will improve faster than we think; however, there will remain some areas with a dose higher than the criteria. This is the fact of nuclear accidents.

We discussed a target number for the return of people, but if we open our eyes to the destruction of life from extended evacuation, we should return the people to their homeland as quickly as possible. The evacuation should be kept in the minimum range for the shortened time to minimize the damage from its influence on people.

Central and local government agencies have to rebuild the infrastructure of society so the returning people can lead regular lives. Autonomous revival of the economy and production cannot be expected unless the population reaches a certain level. Society has to have the correct understanding of "radioactivity" so that not just individual evacuees but the entire local society can make a quick return to its homeland.

Currently, there is no regional map that shows the accumulated dose prediction in a way everyone can understand. We need to map how the regional dose will drop in the future in 3, 30, and 100 years so that people from the area can start planning their future lives.

References

[1] Distribution map of radiation dose around Fukushima Dai-ichi and Dai-Ni NPP, http://www.mext.go.jp/component/english/__icsFiles/afieldfile/2011/08/05/1305904_0720.pdf, 2011, as of July 11.

[2] Emergency radiation exposure medical action procedure (in Japanese), Fukushima Prefecture, May, 1991.

[3] 25 years after the chernobyl accident: summary and overview of its impact and overcoming its aftereffects, 1986-2011, Government of Russia, 2011.

[4] A radiologist's truth about exposure and cancer (in Japanese), Best Paperbacks, 2011, Keiichi Nakagawa.

[5] Tsumagoi, the Bombei of Japan (in Japanese) 1980, Investigation Committee of buried villages at base of Mt. Asama, and Tokyo Newspaper editing office.

[6] Behavior of radiocaesium in soil-plant systems and its controlling factor, National Institute for Agro-Environmental Sciences (NIAES), Bulletin 31, 2012.

Learning from the Fukushima nuclear power plant accident

6

Y. Hatamura
Hatamura Institute for the Advancement of Technology, Tokyo, Japan

6.1 Introduction

The 2011 Fukushima accident was devastating. It badly affected, and is still causing great problems to, the people who lived in the area that surrounded the plant. It has also greatly affected the people of Japan and also those in the nuclear industry around the world. Our lives still go on, and we must clarify what it is that we can learn from this accident.

Accidents happen, and as long as we live in this modern but clumsy world that is full of machines, they will continue to haunt us. As such, we would like to make the best use of these costly experiences obtained in having faced a never-before-seen accident.

This chapter discusses the lessons to be learned from the Fukushima accident; the first section discusses these in detail and the next section considers the valuable lessons that came to mind after having served as the chairman of the Investigation Committee. In fact, these are the same lessons that keep coming back to me every time I am involved with an accident investigation.

6.2 Learning from accidents: key findings from the official investigation

6.2.1 Turning information into knowledge

The Fukushima accident taught us what we have to think about, what to prepare for, and how to manage a plant that produces nuclear energy that has great advantages as an energy source, but at the same time holds enormous risk. The accident gave us directions in the technical field of nuclear power generation, but at the same time, also lessons in all other fields of technology and about how we should face large-scale disasters. We have to take full advantage of lessons learned from this accident.

When we want to learn from an accident, we carry out investigations to clarify what went on and then generate reports. We also often build databases that collect case information. Merely collecting data and analyzing each event on a case-by-case basis hardly makes any use of the data, and they simply fade out from our memory with time.

What is important is to lift our thinking to a higher level of abstraction and generalize the event to turn the information into knowledge. Once elevated to knowledge,

The 2011 Fukushima Nuclear Power Plant Accident. http://dx.doi.org/10.1016/B978-0-08-100118-9.00006-1

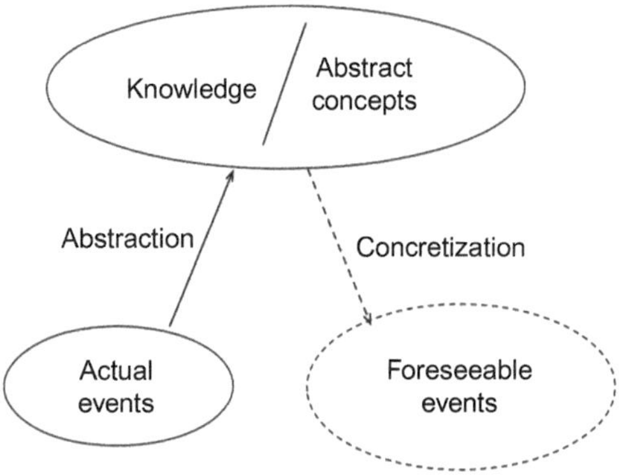

Figure 6.1 Turning failure into knowledge and applying the knowledge. Elevation to abstract concepts and concretization allow foreseeing of events.

we can later apply it to different social situations at the time or new technologies (Figure 6.1).

One example that shows that we did not elevate the lessons from an accident to knowledge and failed to share it within the community is the TEPCO Kashiwazaki-Kariwa Nuclear Power Plant (NPP) case.

The Chuetsu offshore earthquake in 2007 caused damage to the TEPCO Kashiwazaki-Kariwa NPP. The primary facilities of the plant, with plenty of preparation for earthquakes, survived the shaking without major damage. Some of the peripheral facilities, however, suffered damage. Fortunately, the situation was not catastrophic, but damage to the peripheral facilities that had been left without much reinforcement against earthquakes shut the plant down for an extended period. The plant took 2 years to restart, involving not just the rebuilding of the damaged facilities, but also the purchase of alternative fuel burdened TEPCO with unexpected expenditures.

What we should have learned from this disaster is that caring only about primary facilities is insufficient, and we should boost the entire plant, including peripheral facilities, in preparation. Another point we should have recognized is that concentrating all attention in plant safety into just a single event (e.g., in this case, an earthquake) can cause other important matters to slip our minds, and one of those other matters can lead to disaster. The other matter that slipped our minds with the Fukushima accident was the tsunami.

If we had elevated the lessons into knowledge that failure of peripheral facilities, other than the primary ones like the reactor pressure vessel (RPV), can lead to severe accidents (SAs), we would have recognized that not just an earthquake but a tsunami can devastate the safety of the plant as in the case of Fukushima NPP.

6.2.2 Building a total picture with hypotheses

The second important point is to build the total picture of the accident with added hypotheses. This means in addition to tracing the sequence of events (we call this the "scenario"), think of the scenario that would have led to success by adding the

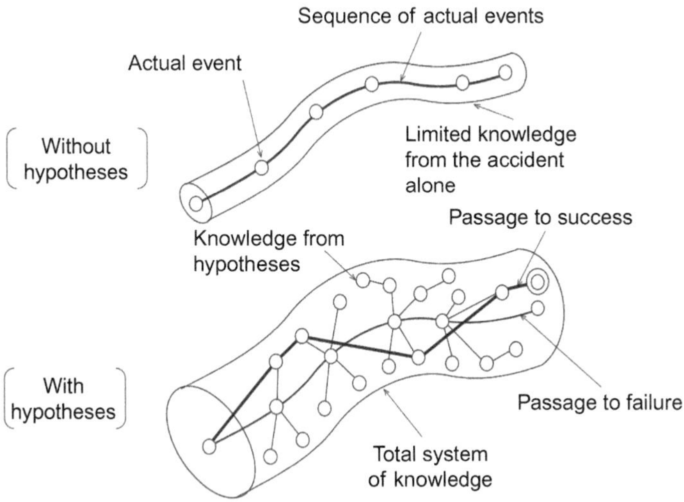

Figure 6.2 Adding hypotheses to the actual events of an accident leads to an enriched system of knowledge.

other choices we could have made in the course of events. Given the actual events, adding such hypotheses builds the real overall picture and leads to an enriched system of knowledge (Figure 6.2).

While events progress amid an accident, we have to make choices and judgments and carry them out as the situations change by the second. The sequence of choices we make during the progress of an accident is the "passage to failure." When we trace the scenario of accidents, we find that we selected and executed routes at each stage that led to failure. On the other hand, if we had made different choices at each stage, we could have avoided the accident, and that is the "passage to success." In many cases, accident investigation analyzes the passage to failure in detail; however, to make use of the lessons from an accident, we have to clarify how we could have avoided the failure (i.e., the passage to success, with hypotheses).

For example, with the Fukushima accident, we found that if at least portable generators and compressors were on hand, as a minimum preparation against station blackouts (SBOs), SAs could have been avoided. Other NPPs in the country immediately took measures to station power-supply vehicles at higher elevations. Such preparations against SAs should have been made before the Fukushima accident.

In addition to making a thorough investigation into the scenario that led to an accident or failure, building the overall picture as knowledge with the search for the path to success allows us to prepare against the next unexpected event. This practice should also let us see what other problems exist at the time.

6.2.3 Thinking after the accident is too late

When people take actions against what is happening in front of their eyes, they consider the options and build plans based on each possibility. Such iterative thinking takes time, and if the event progresses fast, we have to make instant selections and decisions. Thus, we are almost disabled in terms of making the right judgment and action based on logic.

During this accident, a number of events quickly took place one after another, and it seemed following the rapidly developing events with well-thought out plans was impossible. In addition, each choice had to be based on what was visible at the time. Persons in charge often have to make decisions without the overall picture. They have to know that they will be put in such situations and need to make preparations to build the overall picture at all times.

In an emergency situation, like an accident, we can only follow what we had planned for in the past. We have to think of what could happen and how the overall picture would be before the accident takes place. We can then make decisions, plan measures and take action in response to accident situations by finding the most appropriate action from the bag of preparatory exercises.

To plan for all possible situations, however, and be prepared to make the right choice is, in fact, impossible. When we are placed in an unexpected situation, what we depend on are not memories of all possibilities, but the "thinking process" that we built during the preparatory exercises.

We have to bear in mind that we can only perform action according to what we had prepared before the accident. And we have to repeat "hypothetical exercises" during our everyday work so we can build the thinking process within our minds so we can think in the right direction. We should also carry out field training for the time when we have to make the right decisions.

6.3 Recommendations for improving the safe design and operation of NPPs

6.3.1 Properly understanding the accident phenomena: the hydrogen explosion did not spread the radioactive material

Proper understanding of phenomena requires comprehension of the overall picture and at the same time, detailed understanding of each element that constitutes the whole. In other words, we have to capture the overall picture to study the macro-mechanism and at the same time review the micro-mechanism. Many news items and reports about accidents usually cover what stood out in the events and points that are easy for the general public to understand. We then try to understand the situation through just partial information and in many cases end up with the wrong idea about the overall picture.

Many of us have a misunderstanding about the spread of radioactive material in the Fukushima accident. The misconception is "The hydrogen explosions spread the radioactive material that fell on the ground." The scenes of the hydrogen explosions were indeed shocking.

In fact, however, radioactive material was not scattered by the hydrogen explosions, but rather leaked out because the containment vessels (CVs) lost their containment function due the high temperature and pressure caused by the meltdowns.

As we discussed in Chapter 2, Unit 2 that released the largest amount of radioactive material did not have a hydrogen explosion. Its core, however, was damaged and so was its CV, and when it leaked radioactive material to the outside, the external radiation from the entire accident reached its peak.

The question then is, "Which radioactive material made its way to the outside?" Volatile material and light-weighted material spread into the air. Water-soluble material snuck into the water and leaked out from the CV in the form of underground water. Water-resistant walls seem to now be blocking it. However, radioactive underground water had leaked into the ocean during the early stage of the accident (as of April 2014). Insoluble heavy materials are most likely to have remained in the CVs. In short, light materials that are easy to spread or soluble ones made their way to the outside, whereas heavy material stayed within the CV.

The Fukushima accident is often compared with the 1986 Chernobyl nuclear accident. Radioactive material inside blew to the outside during the Chernobyl accident because the reactor vessel itself exploded. The Fukushima nuclear accident, on the other hand, had damage to the RPVs and CVs; however, these vessels did not explode. Radioactive releases to the outside were only those that leaked from the damaged spots. The United Nations Scientific Committee on the Effects of Atomic Radiation (UNSCEAR) announced that it would be publishing two reports in October of 2013. During the press briefing, committee members stated that the amount of radioactive iodine release with Fukushima nuclear accident was estimated to be less than one-third of that with the Chernobyl nuclear accident, and its impact to thyroid glands of residents was estimated to be about 1/60. The briefing concluded that, "It is unlikely to be able to attribute any health effects in the future among the general public and the vast majority of workers." We must recognize that the Chernobyl accident had much bigger external releases, both in terms of quantity and the variety of material.

6.3.2 Proper understanding requires knowing the history

In the 1950s, peaceful use of atomic energy was promoted, and technical development of nuclear power generation started to build the first prototype of a commercial NPP. The government then decided to introduce nuclear power generation as an industry, and the Japan Atomic Power Company (JAPC) started the first commercial nuclear power generation at its Tokai plant. This plant was a graphite-moderated carbon-gas-cooled reactor that is now in the process of decommissioning. Later, instead of gas-cooled reactors, companies built light-water reactors. The construction of GE-designed boiling water reactors (BWRs) with Mark-I containments first started with JAPC's Tsuruga NPP Unit 1 and TEPCO's Fukushima-1 Unit 1. Commercial operations had to wait until 1970 to start (note that KEPCO's Mihama NPP Unit 1, a pressurized water reactor (PWR) design, started operation before Fukushima-1 Unit 1). Remarks: KEPCO: Kansai Electric Power Company.

During the years of 1970-1975, Japan simply licensed GE BWRs. GE designed them, and Japanese manufacturers just contracted the machine, manufacturing, and building construction. Especially in the case of the first Fukushima-1 Unit 1, TEPCO was just the operator. The contract with GE was a turnkey unit.

The technologies introduced then were declared proven. However, in fact, they were still under development with many problems to solve. The Japanese industry, including TEPCO and the manufacturers Toshiba and Hitachi, started serious improvements to the GE-based design. In the second half of the 1970s, Japanese companies with higher skills started their own design and manufacture based on the GE groundwork.

Early nuclear plants had poor efficiency at a level where the utility companies hardly made any profit. TEPCO and the manufacturers Toshiba and Hitachi solved the problems one by one to enhance the quality and improve overall reliability. The efforts gradually raised the efficiency of nuclear plants.

When we introduce technologies from foreign countries, we often have to make partial modifications to adopt the technologies to the new environment and sometimes end up with retrofit designs (designs that use all or parts of designs for other products and fit changes to them) or add-on designs (designs that keep adding changes as requirements change). Once we set the design environment, we should always optimize the overall system. However, whether we followed this practice in introducing nuclear power technology remains questionable.

In fact, natural hazards we have to be aware of in Japan are earthquakes, but in addition, we have tsunami, flooding, and volcano eruptions. When we introduced the American-born technology of nuclear power generation, earthquake preparation was always a topic. Engineers constantly evaluated antiseismic properties; however, they seem to have forgotten the discussion about tsunami risk.

The Three Mile Island (TMI) NPP accident in the United States occurred in 1979. This accident was due to misjudgments by the operators (human error), and it was the first and most dangerous event of a core meltdown.

After this accident, the United States, with the development of combined-cycle fossil-fuel plants, quickly lost interest in nuclear power generation, and although it had licensed about 100 units by that time, no new licenses were issued after the accident until 2012. When no new commercial plants were built for an extended time, the industry lost its solid footing, the manufacturers faded out their development efforts, and although NPPs were being built outside the country, rapid advancement of the technology could no longer be expected.

At the time in Japan, on the other hand, nuclear power generation was promoted, and one to two new plants were built in the country each year. About 50 new nuclear power units were built within these 30 years. Japan showed its high interest in the technology by adopting both BWR and PWR types. Japanese manufacturers put much effort into technology development and made new contributions one after another. This is how the Japanese manufacturers further developed the American-born technology.

6.3.3 No one planned for a huge tsunami

To find what assumptions were made at the time of planning, reviewing the history of Fukushima-1 NPP alone is insufficient. Research about other power plants in the area led us to the answer. We found that when blueprints for building Fukushima-1 were reviewed, almost no consideration was given to possible huge tsunami waves along the Pacific coast south of Sendai. The elevations of nuclear and fossil-fuel power plants damaged by the Great East Japan Earthquake tsunami clearly show the lack of preparation. Figure 6.3 shows the ground elevations of main facilities and the heights of tsunami waves that hit them.

The ground elevation of primary buildings of Fukushima-1 was at 10 m, and that at the ocean side where the seawater pumps were set was 4 m. The tsunami that hit the area was about 15 m high and submerged the entire ground level of the site. Units 5 and 6 of the same Fukushima-1 are located at some distance, and their primary buildings were built on land 13 m above sea level. The tsunami there was about 14 m high.

Fukushima-2 located south of Fukushima-1 was on land 12 m above sea level, and the tsunami there measured 7 m high. The waves, however, ran up to a height of 14 m, and Fukushima-2 was partially submerged.

The Tohoku-EPCO Haramachi fossil-fuel power plant north of the Fukushima plants suffered the greatest tsunami damage. The ground level was at 10 m, and the tsunami there reached 18 m. Main structures of the thermal plant, except the turbine at an elevated floor, were all submerged and destroyed. Nothing had been recovered as of October of 2012 (Unit 2 restarted commercial operation in March 2013).

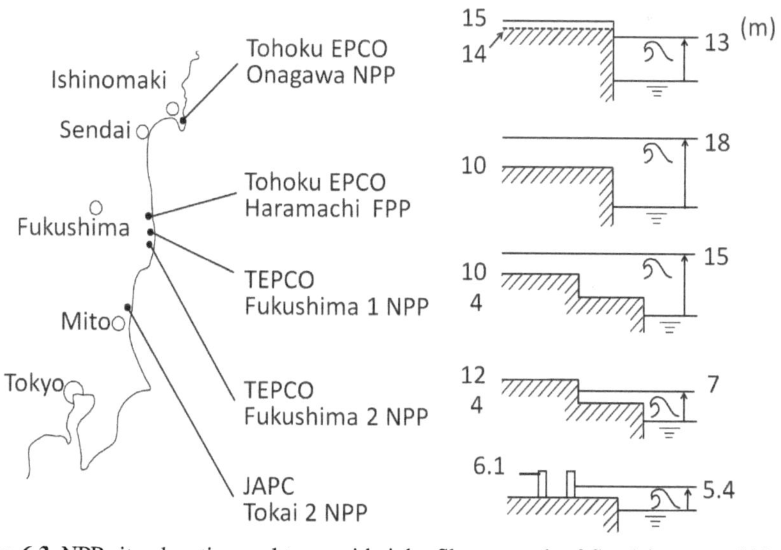

Figure 6.3 NPP site elevation and tsunami height. Shores north of Sendai expected high tsunami, nothing expected in the south. Tohoku EPCO: Tohoku Electric Power Co. Inc.; TEPCO: Tokyo Electric Power Company; JAPC: The Japan Atomic Power Company; NPP: nuclear power plant; FPP: fossil fuel power plant.

JAPC Tokai-2 NPP, 100 km south of Fukushima-1, had seawalls that reached a height of 6.1 m, and the 5.4 m tsunami worked its way in from an unfinished area to stop one of the emergency generators, but the remaining generators supplied sufficient power, and the plant avoided a SA.

Located at the southern end of Sanriku region with frequent tsunami attacks is the Tohoku-EPCO Onagawa NPP. Its site ground level was at 15 m above sea level. The earthquake lowered the ground to a level of 14 m; however, the tsunami height that attacked the area was about 13 m, and thus the plant escaped submergence.

We also need to review power plants other than those hit by the Great East Japan Earthquake. TEPCO Kashiwazaki-Kariwa NPP and Chubu Electric Power Hamaoka NPP both did not have tall seawalls to counter tsunami. This earthquake forced the companies to review their tsunami preparedness, and the plant buildings now have watertight seals and seawalls; they are now reviewing their action plans against SAs.

When TEPCO filed its application to build Fukushima-1 in the 1960s, it set the estimated tsunami height to a maximum of about 3.1 m, a value based on the tsunami from the Great Chilean Earthquake. The technologies back in the 1950s and 1960s had no sophisticated tools for earthquake simulation. However, we must say that there was hardly any consideration of tsunami at the time. The reactor buildings at Fukushima-1 had to be built on a hard rock bed, and the hills, originally at about 30 m above sea, were excavated down to about 10 m, and the soil from the work made the landfill to form the lowest elevation near the ocean at 4 m. The plans had no idea of possible tsunami attacks. Records show that the most recent tsunami preparation had set the estimated tsunami height at 6.1 m, a level that was far lower than the actual 15 m that hit the coast.

From the dawn of nuclear power generation in the 1950s to about 1990, the land of Japan hardly experienced any large-scale earthquake or tsunami. The 1983 Sea of Japan earthquake and the 1993 Southeast earthquake caused some large tsunami waves that hit the Sea of Japan side and left some damage; however, the Pacific Ocean side only had the Great Chilean Earthquake. The tsunami from this 1960 earthquake started offshore of Chile and traveled all the way across the Pacific. Some tsunami preparation after the 1933 Showa Sanriku tsunami was effective against this Chile tsunami to keep the damage to the minimum. This effect led to the unpreparedness on the Pacific side, especially to the south of Sendai.

6.3.4 Japanese NPP safety measures had fallen behind those of other countries

When we look at the accident from the viewpoint of accident management (AM), we realize that the United States took necessary AM measures after learning from the TMI NPP accident.

Japan also took some new AM measures in design and operation guidelines that were found in the United States and were necessary. Connecting the fire-extinguishing lines to the emergency cooling piping was, for example, one of such measures. The United States, after the TMI accident, recognized that the design with no water-level monitors for the PWR pressurizer led to the accident and took preventive measures. In

Japan, the possibility of water-level monitor malfunction with BWRs was identified; however, no alternative ideas for the function merely led to no action being taken.

Moreover, there appears to have been an assumption that levels of education and training in operating power generation systems and the existing skills of field workers meant that accident management planning and execution was sufficient for all emergencies. When we review the actual events with the Fukushima NPP accident, field workers made some wrong judgments and sometimes could not keep up with the progress of the accident.

Until about 1990, Japan had learned from the United States and other foreign countries about troubles with nuclear plants and had been trying to catch up with their countermeasures. The main issue is Japan's response to safety issues from 1990 on.

On September 11, 2001, the World Trade Center of the United States experienced a terrorist attack when airplanes crashed into the two tower buildings. After this attack, the nuclear industry in the United States evaluated what would happen if a similar terror attack was made against a NPP and what measures to take to prevent the attack from leading to a catastrophic accident. The action was a new requirement from the Nuclear Regulatory Commission (NRC) to all licensees known as "B.5.b." The United States communicated its contents to the Nuclear and Industrial Safety Agency (NISA); however, those in charge at NISA did not impose the requirements on Japanese plants.

In 2002, TEPCO's failure to report stress corrosion cracking of one of its plants made the news. A whistleblower informed the authorities that he was forced to hide cracks in the steam dryer. Investigation disclosed that TEPCO executives had hidden more serious cracks in the shroud as well. TEPCO had to shut down all its NPPs for thorough testing. The 1999 JCO criticality accident and the TEPCO incident caused anxiety in the public, and the utility companies started to report their actions to deal with even the most trivial troubles with their plants. NISA had to concentrate resources on coping with all the reports. The situation limited the regulator's ability to review overall accident management planning.

The Great Hanshin-Awaji earthquake occurred in 1995, and then in 2007, the Chuetsu Offshore Earthquake shook TEPCO's Kashiwazaki-Kariwa NPP at a magnitude much larger than the design basis. The primary equipment of the plant hardly received any damage. However, the incident caused earthquakes to catch attention as the worst outside disturbance for NPPs, and engineers concentrated their attention into structural strength. New design standards were set for earthquake preparation, and the plants had to put significant manpower into strengthening their piping support structures and other facilities.

Scientific knowledge about natural disasters like earthquakes, tsunami, flooding, and volcano eruption have made much progress in recent years, and not just nuclear plants, but also the public at large are now required to keep their operations prepared for such disasters.

Before the Fukushima quake, the possibilities of a great earthquake along the Pacific and associated huge tsunami waves were being discussed. However, a respected academic figure had expressed his opinion of the low possibility of an earthquake off the shore of Fukushima. The Sanriku area has had its share of frequent tsunami attacks, but not many people foresaw the possible huge tsunami to the region further south. The risk of huge tsunami had been pointed out based on studies of past large ones like the A.D. 869 Jyogan tsunami. However, TEPCO did not take measures against tsunami on this scale.

While the United States took measures in AM against a wide range of SAs including terrorist attacks, the Japanese industry and regulations focussed mainly on structural strengths against earthquakes. Now that the tsunami has happened, we know that preparations against tsunami should have been made, but no actions were taken before. The earthquake and tsunami hit the land where we were unprepared, and the large-scale disaster was the inevitable consequence. We had to pay a price for not planning against all possible accidents.

6.3.5 Attitudes to nuclear safety in Japan

Nuclear power generation had established its presence as a Japanese energy resource, and the entire public had become dependent on it. Accidents like the TMI accident and the one at Chernobyl led to anxiety in Japan about nuclear accidents, and an antinuclear movement started to spread. Those who were promoting nuclear power generation promoted opinions like "Japan has never had any large accident thus nuclear power itself is not dangerous."

Japan was then faced with the 1999 JCO criticality incident which reinforced public concerns about the possible dangers of nuclear power generation. On some occasions regulators put too much emphasis on alleviating these concerns. As an example, when Niigata prefecture planned an action drill with an assumed nuclear plant accident, NISA questionned the plan for the reason that it would elevate anxiety among the residents. Niigata was forced to downscale the plan to a combined disaster of heavy snow and a NPP accident.

Plants overseas implemented SA measures after the TMI and Chernobyl accidents; however, in Japan these SA measures were not always implemented. An example is the filtered venting systems planned for all nuclear facilities in Europe. Some of the Japanese plants evaluated the new system but they were never implemented. This behavior suggested a degree of complacency about safety improvement.

6.3.6 Safety cannot be assured when promoters regulate themselves

Those that operate NPPs are utility companies like TEPCO, and the Ministry of Economy, Trade and Industry (METI) is the government organization that enforces its policies. NISA was established as part of METI to ensure safety of nuclear power. Those that run a business and the regulators have to be in a relationship that, both structurally and operationally, oppose each other. If a government organization takes responsibility in both promotion and regulation, it is guaranteed to face a conflict of interest.

Before NISA was established, its former organization took charge in the safety of the coal industry. Japan was heavily dependent on coal energy about 40 years ago. The central government reorganization in 2001 formed NISA and put it in charge of safety of the nuclear and other industries including coal. The move was intended to secure safety of nuclear power generation in the same way as with other industries (i.e., to side with the promoters and give advice about necessary safety).

The public expected NISA to supervise the industry to ensure safety by requiring utility companies to prevent all accidents, both large and small. With the public's demand for strict safety with NPPs, NISA, to formalize the establishment of safety, put efforts into tighter inspections and formal reports so it could demonstrate safety to the public.

Instead of building their own inspection lists for NPPs, NISA had utility companies and their contracting manufacturers draft these inspection lists, and then the lists were checked to judge whether safety could be ensured or not. The organization adopted a more reactive role.

When we review the response after the 2007 Chuetsu Offshore Earthquake, NISA single-mindedly went into verifying only the structural strength and checked the material strength of "each and every screw." However, they never reviewed the estimated magnitude of natural disasters and the measures to take against them. That was left as the responsibility of the utility companies. Estimating "what disturbances will take place in what manner," and "what accidents are expected" and taking necessary measures against them was the responsibility of the utility companies. NISA should have taken the role of supervising whether the estimates were reasonable and proper precautions were in place.

6.3.7 *Practicing inverse operation or hypothesis exercise*

We can easily understand that when we are faced with events that exceed the design limits or with something we have never experienced before we can fail to make the right judgment. When the Fukushima accident broke out, a number of problems surfaced. The people that had to respond to the accident lacked a common picture of how the situation would proceed when extraordinary phenomena took place, had incomplete understanding of machines and how the systems operated, had not been trained about the handling of machines in case of emergency, had not laid out action plans to follow. Such problems were not just about the structure of Mark-I containment. When the designer decided to place the RPV inside the CV to prevent radiation from leaking to the outside, he should have planned for all reasonable eventualities.

Here, we will discuss two events, one about a misjudgment during the progress of the accident, and the other about a failure to conduct a planned operation. The first is about an erroneous water-level indication. Upon the accident, the operators believed the water-level indication and misjudged the real water level in the RPV. When they realized their mistake, the core had already melted down. The phenomenon is due to the piping for the RPV water-level measurement passing through the CV. The CV state affects the water inside the piping. The other is about failing to manually operate the air-operated safety relief valve. The backpressure to the cylinder that operates the valve is directly connected to the CV and forcing pressurized air from outside the CV did not open the valve due to abnormally high pressure within the CV.

These phenomena took place because the designer did not thoroughly think about what happens with high temperature and pressure inside the RPV and abnormally high temperature and pressure in the CV. A designer in general may believe he has thought about everything; however, there are always missed events or a slip of mind. We discover them for the first time when accidents take place.

What can we do to predict events that can happen in the future? Many of us rely on our experience and knowledge; however, that is insufficient. A good practice to find important matters is to think what can logically happen using "inverse operation" and "hypothesis exercise" methods (Figure 6.4).

Instead of planning how to make things progress in the right way, inverse operation sees things from the viewpoint of how things can go wrong. That is to assume the accident as a result and trace the scenario backward from the result.

The other method is to think of the progress under different conditions. When we plan things, we think of what will happen, what constraints we have, and what situations we will be placed in to decide the scenario of actions. Hypothesizing what will happen if the conditions were different is what we call "hypothesis exercise."

In practicing inverse operation or hypothesis exercise, we should turn into demons with evil minds and plan how we can make accidents occur. The point is to look at our design seeking insufficient preparations and how we can cause accidents by tweaking conditions.

6.3.8 Assuming the worst-case scenario

Damage from the Fukushima accident grew to a large scale most likely because we had not assumed the worst-case scenario. The prime minister's office hypothesized the worst case, but did not publicize it to avoid unnecessary confusion. Such scenarios,

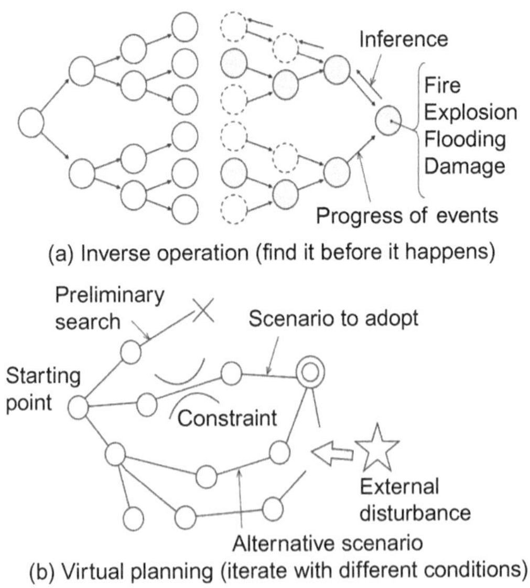

Figure 6.4 Virtual planning and inverse operation for failure prevention.

however, should have been evaluated and made available at the time the plant site was selected and plans were made. Operations should have been carried out with those situations known.

What were the worst-case scenarios with this accident? Studying the sequence of events that lead to such scenarios is one of the most important lessons we can learn from an accident. One of the possible situations was water injection from the fire engines not reaching the cores and fuel pools of Units 1-4. The cores would then melt down to destroy the RPV and CV, and a large amount of radiation would spread to the outside.

The real situation, however, had piping for external water injection leading all the way to the reactor, and the "worst of the worst" scenario was avoided. One of the original AM plans called for possible reactor cooling using the fire-extinguishing pumps in the turbine building. In addition, a fortunate preparation had been made, only 9 months prior to the accident in June of 2010, of opening a water injection inlet on the turbine-building wall.

The plant at the time, however, had not assumed a situation that would make use of these facilities, and there had been no operational procedure or training for using them. Thus, when the accident hit, the preparation took some time, but if the water inlet and piping were not in place there would have been no way of injecting water from the outside.

The other possible worst-case scenario was destruction of the Unit 4 fuel pool and the huge radiation leakage it would have caused. Although the hydrogen explosion blew away the upper part of the Unit 4 reactor building, the pool survived the blast. Also, the unit was in periodic maintenance, and the water in the reactor well naturally made its way to the fuel pool. If the situation had been different, the large amount of fuel stored in the fuel pool would have overheated leading to a huge radiation leakage to the outside. Academia must further verify the situation that the world feared the most.

If such a worst-case scenario had broken out, everyone on-site attempting to control the plant would have had to evacuate the area, and Fukushima-1 would have been left on its own. A huge amount of radiation would have spread over eastern Japan, and the daytime population of 30 million in Tokyo would have had to evacuate the area.

One thing that we should not forget is the activities of the workers in the plant field who responded to the changing situations at the risk of their own lives. What we have learned is the importance for plant owners and all related persons to assume the worst case and make sufficient preparations against such situations. Each and every worker on-site should be aware of such hypothetical situations.

6.3.9 Focusing attention on the wrong aspects

When a person thinks hard, his thought forms a hierarchical structure, and usually the person concentrates his attention on the most important part. However, when his work environment demands formality, he starts to spare his attention to the lower levels of the hierarchy. In the end, he fails to turn his attention to the most important upper levels, and an accident breaks out (Figure 6.5). This is what happened between the utility company and NISA. This problem of a shift in the range of attention is often observed in mature societies.

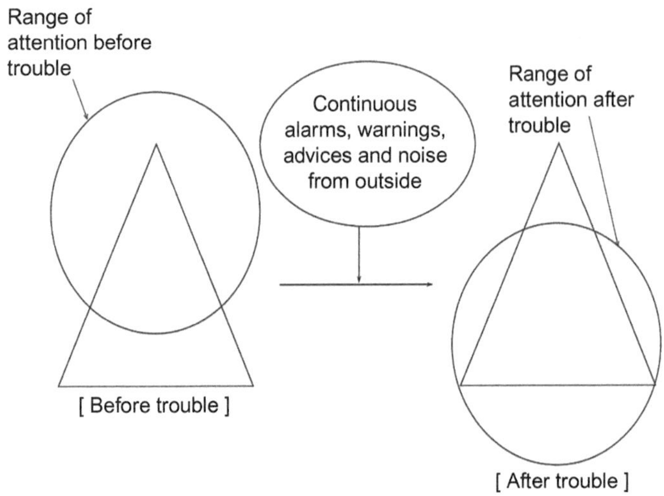

Figure 6.5 Limitations to people's attention.

The regulating body NISA concentrated its efforts on enforcing existing regulations. Those being regulated had to prepare significant amounts of data and formal documents. This made it harder for both regulator and companies to focus on wider safety issues.

In contrast, the U.S. NRC has those in charge stationed on-site to access all documents and walk inside the plants anytime for inspection. Instead of requesting formal documents from the plant employees, they would walk inside the plant themselves to make their own judgments in filling out their own forms. This more proactive approach makes it easier to identify potential safety issues not covered by regulations.

The lack of AM due to concentrating on matters like the strength of material of each and every screw was caused by exaggerated attention to earthquakes. The excessive concern with structural strength led people to lose the overall picture. This is evidence of people's attention shifting down from the top of the triangular hierarchy to the bottom.

6.3.10 *Emergency situations require special measures*

National and local governments often prepare standards and manuals that try to cope with nuclear accidents with manners under normal situations. If the standards and manuals had properly assumed emergency situations when they were made, they would have been able to cope with such situations. This time, however, under rapid transitions with accidents or with the outbreak of the unexpected, efforts to stick with the first set measures led the people to get lost in the events.

What happened in instructing people to take stable iodine tablets, the incident we discussed in Chapters 3 and 5, was one such example of the authorities failing to take proper actions for following guidelines made for normal circumstances. Under normal circumstances, the process of giving instructions to the residents to take stable iodine tablets starts when the Nuclear Safety Commission (NSC) of the government advises the administration to the medical team of the local NERHQ. This committee then makes decisions for prophylaxis administration and reports it to central NERHQ. The headquarters then instructs the administration, and this decision is delivered to the regional emergency headquarters and then to each prefecture governor, who relays the decision to the residents.

Stable iodine tablets, however, are meaningless unless taken before radioactive iodine is inhaled. Thus, if there is a threat of radioactive iodine release, the administration should immediately be instructed. Sticking with the rules of stable iodine tablet administration cannot lead to a quick and effective prophylaxis administration.

On the other hand, there were local governments like Miharu that instructed residents to take stable iodine tablets that had already been distributed. The prefecture of Fukushima was concerned when the town of Miharu instructed distribution and intake of stable iodine tablets without the instruction from higher authorities. The prefecture ordered Miharu to cancel the instruction and recall the tablets. The action by Fukushima prefecture was inappropriate in the circumstances of the emergency.

The Fukushima accident taught us that placing priority on the formalities of regular situations and bounding ourselves to them could cause wrong judgments about the seriousness of events and lead us to fall behind the disaster's progress. There are times that proper measures can only be taken at the right time, and the right time may never reach us again.

So, how should we cope with an emergency situation? First, we have to plan against all possible emergency situations and prepare to make the proper selection. It is necessary to set out measures to take with the assumption that situations progress rapidly at the time of the emergency.

At the same time, we need to prepare against situations that we had not planned for. Each member of a group should exercise hypothetical planning to build proper tracks of thoughts within the mind and establish priorities of value that serve as the basis for judgments. The entire group then has to share the same priorities of value.

A basic rule in operating an organization should distinguish between an emergency and a situation. When the chief officer declares a state of emergency, operations should immediately switch to such operations that allow delegation of authorities with different channels of command.

6.3.11 Learning from other industries

The utility companies, administration, academia (to provide engineering support), and the related manufacturing industry formed a group focussed on nuclear power generation. This has been described as the so-called nuclear village. A danger of such groups is the failure to learn from those outside the group.

As an example, the 2007 Chuetsu Offshore Earthquake triggered a fire from a transformer in Kashiwazaki-Kariwa NPP, and the underground water piping broke and could not deliver water to the fire-extinguishing equipment inside the plant. Chemical plants, on the other hand, usually have their fire piping aboveground because underground piping breaks with earthquakes and the damage cannot be seen. This is an example that the nuclear industry did not learn from other industries and never reviewed and reevaluated their design methodology of categorizing equipment in only the three classes that they had prepared.

The authors studied the history of boilers, one of the base technologies that contributed to the development of industries after the eighteenth-century industrial revolution. We reached a hypothesis that "Accumulating enough failure experience within a single industry field takes 200 years."

The boiler was invented in the eighteenth century and established its practicality in early nineteenth century. As it took higher pressure, the industry faced explosions with many casualties but enhanced its safety by adding new safety standards. After the technologies for material control and welding matured, the U.S. American Society of Mechanical Engineers (ASME) lowered its standard's factor of safety from 5 to 4. When we plotted the history on a curve, we found that railway, aerospace, automobile and other industries followed the same trend (Figure 6.6).

The nuclear power generation industry, on the other hand, started applying nuclear power to electricity generation in about 1950, and since then, only 60 years have passed. Major accidents during these years were TMI, Chernobyl, and Fukushima. The direct causes of the accidents were human error in case of TMI, a design flaw that missed self-controllability in the reactor for Chernobyl, and lack of preparedness against the natural disasters of earthquake and tsunami in the case of Fukushima.

So far, we have experienced these accidents; however, further major accidents may still take place. They can be caused by terrorism, a person or a group of them with twisted mind, "coincidence" of multiple happenings, and so on. Reaching a state of assured safety by following standards, guidelines, past experience, and knowledge will take another 140 years if we follow our preceding hypothesis.

We believe, however, that if the field of nuclear engineering learns what happened with other industries and internalizes the knowledge, it can shorten the 200 years term (Figure 6.7). That is, if the nuclear industry thoroughly studies knowledge from the

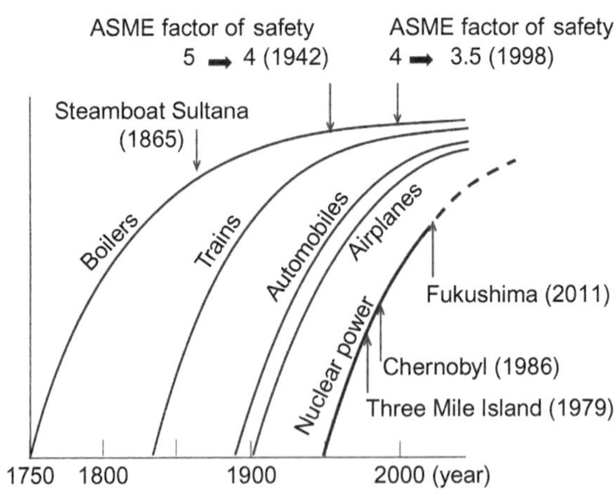

Figure 6.6 Any industry takes 200 years to gain enough experience of failures. Nuclear power is only 60 years old.

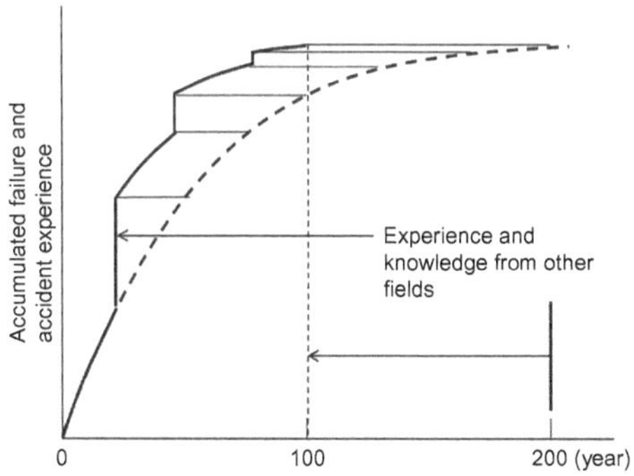

Figure 6.7 Learning from experience and knowledge in other fields allows shortening the learning process.

other fields, with great effort, maybe it can shorten the remaining learning period to 40 years to reach a state similar to other industries, or perhaps even a higher level for nuclear power generation.

6.4 Comparing the Fukushima accident with other nuclear incidents

In this section, we would like to assess whether, if we had acquired knowledge by learning from past accidents before the one with Fukushima NPP, we could have prevented this accident (disaster prevention)? And could we have minimized the damage (disaster reduction)?

This line of thought comes from Section 6.2.2.

6.4.1 Three Mile Island (TMI) NPP, USA, accident

The 1979 accident in TMI in the United States was caused by a stuck-open pilot-operated relief valve. The primary system cooling water leaked from the valve to expose the fuel that melted down. This was the first major accident that taught us about "human error" that no matter how properly the system may be designed, a human operator can make a misjudgment or perform an erroneous operation.

This accident taught us a number of lessons and afterward, many countries in Europe installed a filtered CV vent in case the system pressure rises too high; however, Japan did not take such precautions.

Another lesson learned with the TMI accident was about the water-level indication of the primary cooling water system for the PWR. The design at the time did not have one, and if there was, the operator would probably not have made the misjudgment. Fukushima-1 reactors were all BWRs with water-level indicators; however, they had known reliability problems. No measures had been taken against the problem, and when the accident took place, the operators failed to recognize the erroneous water-level indication and let the accident progress into a worse situation.

Also, in case of the TMI accident, an off-site person, with an overall picture of nuclear power reactors, recognized the possibility of a stuck valve or wrong operation, and his pointing out the possible problem turned around the situation to stop the runaway reactor. The person with an overall picture with detailed knowledge was the key in preventing further expansion of the accident. On the other hand, with the Fukushima NPP accident, there was hardly anyone with an overall picture and detailed knowledge, and the accident kept making unwanted progressions. It was not just the problem with the water-level indication, but those who should have had a picture in their minds about the overall situations of the accident had to make judgments based on local information. This fact reveals that the plant had problems with personnel arrangements for sharing fundamental ideas about the plant.

6.4.2 Chernobyl NPP, former Ukrainian SSR, Soviet Union, accident

In 1986, an accident took place in Chernobyl in the Soviet Union. This nuclear reactor was a carbon moderated light-water cooled boiling reactor, a reactor type unique to the Soviet Union with a totally different structure from a BWR or PWR. During an emergency cooling system test, the reactor turned unstable, and the operator inserted the control rods. This insertion caused a sudden rise in the reactor power that resulted in an explosion and fire. The Japanese industry described this accident as one with a nuclear reactor with divergent control system saying not much could be learned from it.

This Chernobyl NPP accident forced over 130,000 residents within 30 km from the plant to evacuate. The biggest eye-opening statistic with this evacuation was the follow-up study that showed those that fled the area lived an average life 7 years shorter than those that remained in the highly radioactive area. People cannot tolerate a sudden and drastic change in lifestyle, and that effect is much greater than health-affecting damage from radiation.

In Japan, no evacuation plan had been formed against a realistic hypothetical accident or an evacuation drill in a form close to reality. This lack of preparation caused, when the real one hit, an evacuation of 160,000 residents who are now (as of March 2014) still evacuated from their homes. During a single year around 2013, there were about 1.5 earthquake-related deaths per day among the evacuees. Maybe not all were caused by the nuclear accident, but many of them were those evacuated because of the nuclear plant accident. We have learned, from the Chernobyl accident, that such a drastic change in lifestyle has a much greater influence on people; however, Japan gave no consideration to that fact. If we had properly learned from the Chernobyl accident, we would have recognized the need for preparing proper evacuation plans, carrying out evacuation drills, and sharing the right knowledge of the effects of radiation to the human body and the environment. We would also have judged how to carry out the evacuation by balancing the effect of radiation on human health and that of changes in life environment. The schedule would have been completely different from the current evacuation in Japan, and probably the tragedy of 1.5 daily deaths would have been lessened.

6.4.3 JCO, Japan: criticality and exposure accident

In 1999, a criticality and exposure accident took place at JCO in Japan. Workers at JCO were reprocessing spent uranium fuel when they deviated from the standard procedure that was too slow to meet the required turnaround in handling concentrated uranium solution. The workers, who were aware of the danger of criticality, had their own ways of reprocessing the fuel. When another worker got the job from workforce rearrangement, he was unaware that the process was violating the standard procedure and even made further simplification for added speed to cause criticality.

The direct cause of this accident was a deviation from the process manual; however, in the background was the fact that the workers were subcontractors charged with lowering the cost of production and had to meet an impossible turnaround time. The company at the time needed some reorganization. And the reorganization, which seemed rational but in fact was just replacing good management with one that lacked a real management mind, resulted in a work procedure that deviated from the manual. The accident was a result of management that did not take real responsibility and that only followed formalism.

Meanwhile, Japan, after this simple accident enacted and enforced the "Act on Special Measures concerning Nuclear Emergency Preparedness." Unfortunately, the law was helpless when faced with the compounded disaster of earthquake, tsunami, and nuclear accident all happening at the same time. The fact is the authorities should have set the law to counter even the most complex accident event where multiple factors intertwine with one another, and that was a possible situation.

Here is an example. After this accident, an off-site center was built for each and every nuclear plant in Japan. Its purpose was not properly understood, and it was not properly prepared. The off-site center for the Fukushima plant did not function at all during the Fukushima nuclear accident because a radioactivity filter, although it had been budgeted for, had not been installed. The people that gathered at the Fukushima off-site center had to face the embarrassment of fleeing all the way to the city of Fukushima before taking action. When we establish a new law, if the people required to take action or make decisions to its effect do not share the spirit of the law, the result is an incoherent meaningless act.

6.4.4 Le Blayais NPP, France: water submergence event

In 1999, Le Blayais was partially submerged and faced an emergency situation. The submergence was caused by the flooding of the Geronde river that runs by the plant. Severe wind and rain triggered a water-level rise, and high waves in the river flooded the pumps and electricity facilities to stop the electric supply. Fortunately with this accident, other cooling systems operated, and power was recovered in a relatively short time to avoid the state of emergency. Lessons that should have been learned from this accident are that a nuclear plant could face water submergence, and preparations should be made to supply electricity and keep core cooling running even in such cases.

Japanese industry and regulators did not recognize the significance of this accident and did not plan against water submergence as an external factor that could lead to a severe accident. If people had evaluated water submergence as an external factor, they would have naturally planned against tsunami, and proper preparations could have been put in place.

6.4.5 Maanshan NPP, Taiwan: loss of all AC power event

In Maanshan NPP in Taiwan, an accident took place in 2001 where the plant lost all its AC power supply. Salt damage crippled the power-supply system forcing the plant into a SBO. The plant was equipped with portable emergency diesel generators as a preparation against SBO. Hooking the generator up quickly put an end to the SBO. Lessons to learn from this accident were that even when the generators and off-site power were normal, troubles with the power bus or the switchboards (on-site failures) could cause a SBO, and that in such situations, portable emergency diesel generators prove extremely valuable. Japan, however, confident about the high reliability of off-site power in Japan and about having inter-unit power sharing in place did not learn the lessons from this accident in Taiwan.

6.4.6 Multiple simultaneous terror attacks in the USA

In 2001, a multiple simultaneous terror attack took place in the United States. This incident taught us that an evil attack targeting a NPP is possible in addition to natural phenomena or human error. After the attack, the U.S. NRC issued an order called "B.5.b" to prepare AM measures in case of loss of AC power, maintaining cooling of spent fuel pool, and so on. Japan, however, did not learn from this order in the United States and made no additional preparation against such possible events. If the Fukushima plants had the most basic preparations in B.5.b of emergency generators and compressors, the Fukushima NPP accident would likely have been prevented.

6.4.7 Kalpakkam NPP, India: effect of Sumatra-Andaman earthquake and tsunami

In 2004, the Sumatra-Andaman earthquake and tsunami it induced stopped the Kalpakkam NPP in India. Tsunami waves from the earthquake traveled across the Bay of Bengal to reach India. The NPP pumps were inundated, and an emergency shutdown of the reactor took place. IAEA guidelines had required nuclear plants constructed at locations with possible flooding to provide watertightness in addition to constructing levees and banks. In Japan, in contrast, such measures were hardly evaluated. Especially preparing watertight seals for buildings was never addressed. If Fukushima NPP had its buildings retrofitted for watertightness, the Fukushima accident would not have been so catastrophic.

Lessons we should have learned from the Sumatra-Andaman earthquake were not limited to the countermeasures against flooding, but there was far more to learn with regard to "natural disasters and assumptions." Fukushima NPP had based the intensity levels of natural disasters on its own past experiences, and hopeful boundaries and measures were made. No preparation was made to counter earthquakes with magnitude 9 or greater, and neither was there any preparation against associated tsunami. The lesson from the Sumatra-Andaman earthquake was earthquakes and induced tsunami can reach extremely large magnitudes when they are ocean trench types. Japan, especially with its geographic location, is always exposed to the risk of ocean trench type earthquakes, and evaluations should have been made after the Sumatra-Andaman earthquake.

6.4.8 Kashiwazaki-Kariwa NPP, Japan: emergency shutdown during the Chuetsu offshore earthquake

In 2007, the Chuetsu Offshore earthquake forced an emergency shutdown of Kashiwazaki-Kariwa NPP. The acceleration from the earthquake reached a maximum value close to 1000 Gal; however, no severe incident such as meltdown took place, and primary facilities like the reactor vessels or electricity supply systems hardly suffered any damage. The earthquake did not trigger any tsunami, leading everyone in the nuclear industry to conclude that planning against earthquakes would be sufficient. This Kashiwazaki-Kariwa incident was taken as an example of the success of nuclear plant preparedness. Everyone was concerned about the primary facilities, but the secondary ones that surrounded the primaries suffered some damage; for example, a fire from an auxiliary power supply short circuit and failure to extinguish the fire due to broken fire protection piping. This damage in the secondary facilities blocked the restart of the plant for 2 years.

This was an accident where the power plant failed to restart due to accidents with secondary equipment that surrounded the primary facilities. This accident, in fact, was teaching us the significance of minor accidents in the peripherals working their ways toward the center to cause definite damage to the function of the primary facilities. This is the viewpoint of "peripheral damage" that can affect the overall function.

In contrast, there was one example of a lesson learned that gave positive results with the Fukushima NPP accident, and that was the Seismic Isolated Building. People recognized the need for this building through lessons from the Chuetsu Offshore earthquake, and Fukushima NPP had one built. If this seismic isolated building had not been there in the Fukushima NPP, there would have been no central location for disaster response, and the worst possible situation could have been taking action

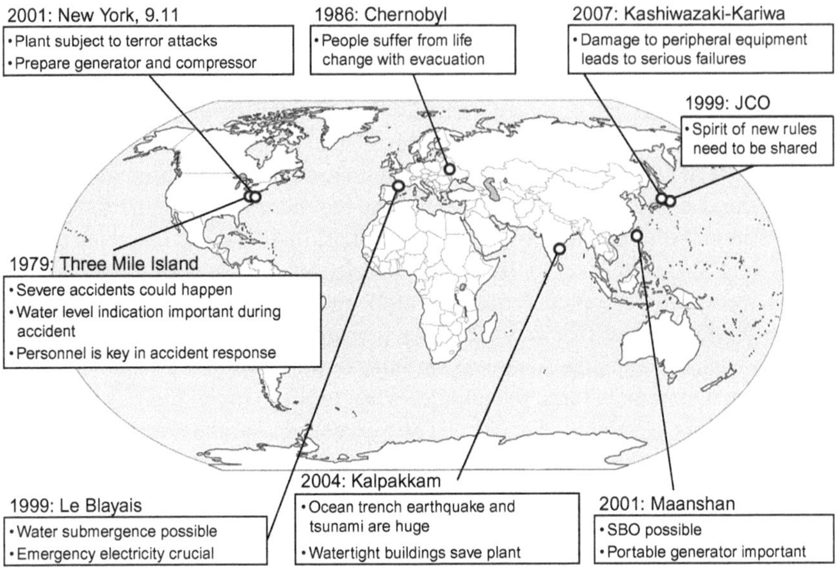

Figure 6.8 Lost opportunities of failure and successful lessons for Fukushima-1.

while being exposed to high radiation. The fact that this building was standing in place was probably the single successful lesson from the Chuetsu Offshore earthquake.

The preceding paragraphs picked out some accidents that took place before the Fukushima NPP accident. Analyzing them would have taught us what to do to prevent accidents (disaster prevention) and how to minimize damage from accidents that have occurred (disaster reduction). Figure 6.8 shows these events with their locations over a world map. We also discussed a number of hypothetical situations of what would have happened if we had learned from the lessons and had taken measures beforehand. It is now Fukushima NPP accident's turn to teach us what knowledge we have to share, and that is to build hypotheses to find out what we should have learned from past accidents.

6.5 Reflections from the Chairman of the official Investigation Committee on the accident at the Fukushima NPP

I, Hatamura, one of the authors of this book, wrote "Afterthoughts by the Chairman (Chairperson's Remarks)" in the Final Report by the government investigation committee [1]. What I wrote was the knowledge of what we learned from this accident. When I first accepted the assignment to be chairman of the investigation committee, I set my goal of producing an accident report that will still have high value 100 years later. We must turn what we learned from this accident into knowledge, and the knowledge must be generalized so that it will be applicable to various fields even after time has passed. I here quote the seven afterthoughts as they appeared in the final report.

(a) What could happen happens, and what we think could not happen happens as well.
(b) We do not see what we do not want to and only see what we want to.
(c) Expect to the fullest extent possible and make the fullest preparation.
(d) Forms alone do not function. They may provide mechanisms but do not share the purpose.
(e) Everything changes over time. We have to stay flexible against changes.
(f) We need to cultivate a culture that recognizes risks and squarely faces them to discuss their management.
(g) Admit the importance of seeing for oneself, thinking individually, making own judgments, and acting accordingly. We need to nourish such capabilities for ourselves.

Of the seven points, (a), (b), and (e) are about how to see things and think. Points (c) and (d) are lessons for organizations, (f) is about culture, and (g) is for individuals. The rest of this chapter will look at each point in detail by reviewing events during the accident that led me to reach the conclusions.

6.5.1 How to see things and think

What could happen happens, and what we think could not happen happens as well (a)

In 1977, the Japan Atomic Energy Commission (JAEC) reviewed "Guidelines for Evaluating Safety Design of Light Water Nuclear Power Reactor Facilities," and concluded that "··· electrical power-supply systems have high reliability and there is no need to evaluate their loss of function." Furthermore, in 1993, NSC denied the need to evaluate an extended loss of AC power supply.

The overconfidence in the quality of electrical power supply in Japan is probably responsible for these guidelines, and the guidelines led people in the Japanese nuclear industry to think that preparation or training for an extended loss of power was not necessary.

When people turn their attention to frequent tedious troubles, they tend to forget about accidents that lead to serious consequences but with low frequencies (Figure 6.9). Think, "What could happen happens, and what we think could not happen happens as well," and study accidents and experiences that took place within the country and overseas, plan all factors to find all logically possible events, and at the same time, take at least all measures that are absolutely necessary.

Especially for NPPs, that can cause catastrophic damages in case of SAs, looking at events that take place inside the plants is insufficient, and we need to assume serious damage to the reactor core caused by natural disasters that greatly surpass the design standards or terror attacks. We need to evaluate and prepare effective measures against SAs.

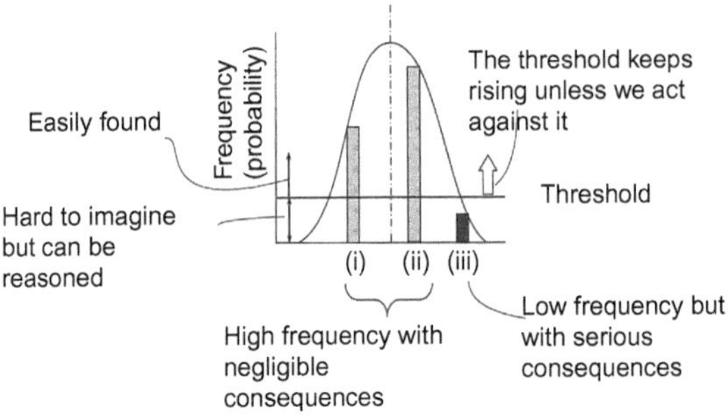

Figure 6.9 What could happen happens. Things unlikely to happen and even events beyond imagination happen as well.

We do not see what we do not want to and only see what we want to (b)

When we see and think about events that take place, we are affected by our own interest, organization, society, and tend to see only what we want to see and not see what we do not want to or what can cause us negative effects.

As we discussed in Section 6.3.3, one of the reasons for the overly low estimate of tsunami height at Fukushima-1 was a general underestimation of the level of risk from tsunamis. People's awareness of tsunami varied along the Pacific coast of the Tohoku region. People in the Sanriku area north of Ojika Peninsula, with memories of Meiji Sanriku Great Tsunami and Showa Sanriku Great Tsunami, were highly cautious about tsunami. Those in Sendai bay and to its south, on the other hand, were barely alert to their possibility. The Great East Japan Earthquake revealed the fact with the high fatality count in Ishinomaki that faces the north side of Sendai bay. The regions around Fukushima NPPs were unprepared against tsunami as well.

TEPCO's lack of preparedness, despite the possibile risk, reflected this broader view of the level of tsunami risk. In addition many seismologists believed "offshore of Fukushima is free of great earthquakes." The company and others fell into the trap of "not seeing what it did not want to."

For others not to fall into the same trap, we need to know our tendency of looking away from inconvenient facts, acknowledge that we always have oversights, and have the attitude to search for phenomena that we do not want to see, inconveniences, and what we do not want to happen. Figure 6.10 shows that when we do not want to see them, we cannot see the demons of risk hiding behind the trees, but once we try to find "what we do not want to," we can locate the hidden demons.

Expect to the extent possible and make the fullest preparation (c)

Fukushima NPP had rigorous preparation against earthquakes, and primary facilities probably did not suffer any damage. Its preparation against tsunami, however, was poor. If it had equipped itself against the unexpected, the accident most likely would not have turned into such a catastrophe.

Figure 6.10 People do not see what they do not want to. Different viewpoints can uncover dangers.

"Expectation to the fullest extent possible with sufficient preparation" may sound like asking for an infinite solution, and a seawall that would withstand it would require a limitless cost and is practically impossible.

My intention here is for measures that do not let unexpected situations lead to the worst-case scenario. For example, watertight doors on buildings and emergency portable power supplies and compressors would have prevented the severe situation caused by this accident.

We, on the other hand, have to admit that we always leave events that we fail to recognize (Figure 6.11). We have to accept the possibility that "unexpected things happen" and build preparations that do not lead to the worst situations. This means that we have to humbly admit that human wisdom is limited.

What we need at times like this is "disaster reduction" instead of "disaster prevention." Reducing disasters means not only taking measures to avoid damage but also preparing measures that can minimize the damage even when disasters break out.

The safety myth led us to think that we had absolute nuclear safety, and we failed to prepare ways of reducing damage in hypothetical accidents. Whether an accident or natural disaster, it will be ideal if we can fully prevent their outbreak. However, this by all means is not an indication that we do not need to prepare measures for disaster reduction in case it strikes.

Everything changes over time. We have to stay flexible against changes (e)

Our findings, knowledge, environment, and social opinions all change over time. We shall never fix these matters and stay flexible in search of appropriate actions.

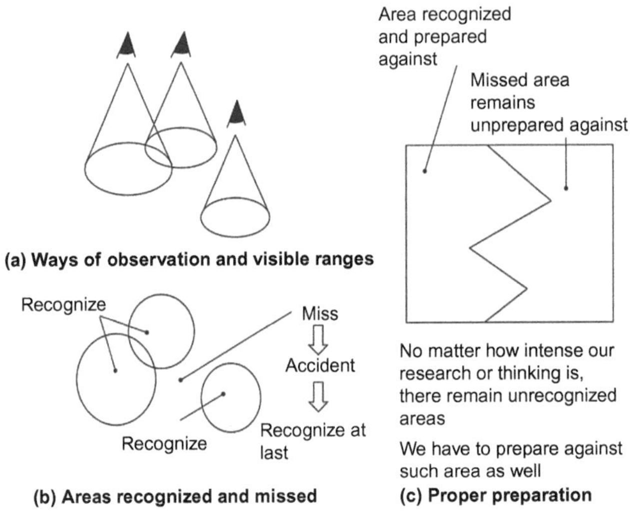

Figure 6.11 Even the hardest thinking leaves unrecognized areas.

TEPCO, for example, as we stated earlier, may not have had sufficient knowledge at the time it prepared Fukushima-1 NPP. It, however, failed to review the possibility of great tsunami to the Pacific coast of Fukushima prefecture even with the advancement of seismology and research about the Jyogan tsunami.

When we turn toward past events overseas, we find nuclear plant accidents at TMI and Chernobyl, SBO at Maanshan NPP, SBO from flooding at Blayais NPP, and also the raised concern about nuclear plant safety after the 9/11 terrorist attack to the United States. TEPCO, however, failed to review its preparedness.

In fact, we can see how people change their opinions over time if we follow how they changed after the TMI nuclear accident in the United States. The TMI accident took place in 1979. The United States had no plans for building new nuclear plants after that. However, 30 years later in 2012, a license was approved to build a new NPP.

When we think about nuclear power generation in the years to come, we have to realize that the public changes its attitude toward nuclear power generation over time. Otherwise, we can easily make the wrong judgment.

6.5.2 *Lessons for organizations*

Forms alone do not function. They may provide mechanisms but do not share the purpose (d)

When we build an organization, without its members fully understanding what its purpose is and what the public expects from it, it will fail to meet the original required function. Also, when we build systems or facilities, if those involved with them do not understand the underlying mechanisms, they will not be used to the fullest capacities.

Chapter 5 discussed the failure to make use of information from SPEEDI because the organization in charge believed accurate prediction relies on release data, and even though the terrain and weather information predicted how the radiation would spread, the information was never made publicly available. The evacuees had no idea other than trying to move away from the nuclear plant, and many of them ran in the direction the radiation was spreading to. This situation can only be blamed on those in charge of SPEEDI not fully understanding its purpose.

Also, as we pointed out in Chapter 3, the Off-site Center, which was designed to serve as the local emergency response headquarters located 5 km away, never served the purpose, despite its budget for facilities for radiation protection because it had taken no action for 3 years. The response headquarters function was forced to move to the Fukushima prefecture government building at a remote location. This is an example of something not performing its required function even if it was shaped to the fullest form.

When we review this accident and its course, we find plenty of such examples. Nobody believed in such a serious accident, and all measures that had been prepared were merely made to meet the formality requirements.

If we had assumed that SAs could take place, we would have found the shortcomings of our prepared measures and would have planned what we could do in case of such a situation. A complete plan on the desk or simulated actions always has something missing, and we never recognize them until we make real actions. One of the root causes for having great forms but failing to meet its function is in the organizational structure and people that are part of it.

Most organizations have sectionalism. The vertical structures work effectively well at regular times. Such structures, however, stop functioning at emergency situations with rapidly changing conditions and new problems to solve every second. At times of emergency, lateral coordination is often most important. Organizations must operate by removing the walls among vertical structures and cooperating in a lateral manner. We have to plan ahead of time about when, how, and under what situations we should switch to lateral coordination mode. Starting to build plans after things have happened is too late.

During this accident, serious events took place one after another and quickly changed the situation. And to make things worse, there were no real plans about organizational operations at times of emergency. When we look at it now, our operations within organizations totally lacked sense.

This problem of sectionalism is not limited to the central and local governments and the utility company. We can probably say it is a common problem to all organizations in Japan. We have to look at this problem as a widespread disease throughout Japan instead of seeing it as merely an issue with SPPEDI, the Off-site Center, its members, or its organization.

The head of an organization has to make final decisions about all matters that take place. In this case, Prime Minister Kan had the role for the country, and Plant Manager Yoshida had the responsibility for Fukushima-1.

When the Fukushima nuclear accident broke out, Kan had to think and make decisions about a number of things. The organizations had the important roles of supporting his analyses and judgments. The reality was, however, there was hardly anyone to answer questions that Kan had or to give him proper advice.

The situation forced the prime minister himself to fly to Fukushima NPP by helicopter. This action was later criticized; however, such blaming of actions later are out of place. The real problem was the lack of proper support when the person at the top had to meet his responsibilities.

In the case of Fukushima-1 Plant Manger Yoshida, he had to act against the worsening situation in the field second by second while not receiving sufficient support from TEPCO headquarters. When TEPCO headquarters requested to stop injecting seawater as Prime Minister Kan wanted, Yoshida, out of fear that the stop would worsen the situation in the reactor core, pretended that he gave orders for stopping seawater injection, but in fact continued it. It was a great performance by Yoshida, who was well aware that things do not happen unless the formalities are followed.

We can see the problems of Kan and Yoshida in having to make critical judgments and execute them while the organizations were not at all fully functional. We have to recognize the fundamental social problems in these situations and make efforts so things can approach their ideal state.

6.5.3 Cultivating the right culture

Cultivate a culture that recognizes risks and squarely faces them to discuss their management (f)

When there are both pros and cons against a matter, the Japanese culture avoids confrontation of the opposing parties and likes to see an intermediate resolution.

In terms of nuclear power generation, pros and cons were in conflict, but the Japanese culture that avoids debate failed on both sides to recognize the risks with nuclear power and stopped planning against them.

Nuclear is extremely efficient as a source of power. It, however, always comes with the risk of release of radioactivity and exposure to radiation. In addition, it is a technology that has not reached maturity (e.g., the disposal of spent fuel has not yet been established). When operating a NPP under such circumstances, we have to review the balance between convenience and risk and discuss how high a risk we would accept in using the technology. Regrettably, however, we never discussed such matters in Japan and in response to the call for nuclear power generation, headed in the direction of promoting it.

We cannot deny that our demand for absolute safety and ease of mind had led to a lack of disaster reduction measures and poor preparation in preventing the spread of damage.

"Safety" means a state that is free of accidents. The ideal situation, however, of absolute safety is impossible. There cannot be a state of no risk.

It may be natural for humans to seek ease of mind. In common sense, "ease of mind" means a state in which we do not have to think about risks. Not thinking about risks, however, leads to accidents. Then, how can we reach real peace of mind? It is a state where we understand the nature of the risks, take measures so they do not arise, and even if they did we have preparations in place to minimize the damage they cause. For us to reach such a state, we must squarely face risks, without looking away from them as "something bad," and recognize risks as risks.

Before this accident, both pro- and antinuclear parties were the same in that they were after safety (Figure 6.12). Pro-nuclear parties claimed that nuclear power is safe because measures against expected risks have been in place, and antinuclear parties argued not to use them because the measures were insufficient or the starting points were wrong. Both parties lacked the viewpoint that accidents happen, and what we really needed back then were measures to minimize the damage in such cases.

We have to admit that no matter how hard we think, we always miss some points, and we must make necessary preparations against possible accidents. If we can maintain a state of preparedness at all times, that will be the real safety for us.

6.5.4 Thinking as individuals

Admit the importance of seeing for oneself, thinking for oneself, making one's own judgments, and acting accordingly. We need to nourish such capabilities for ourselves (g)

The Fukushima accident was severe with three nuclear reactors reaching core damage one after another, but the current level of damage was the result of field efforts that prevented the worst situation. The Fukushima plant workers risked their lives

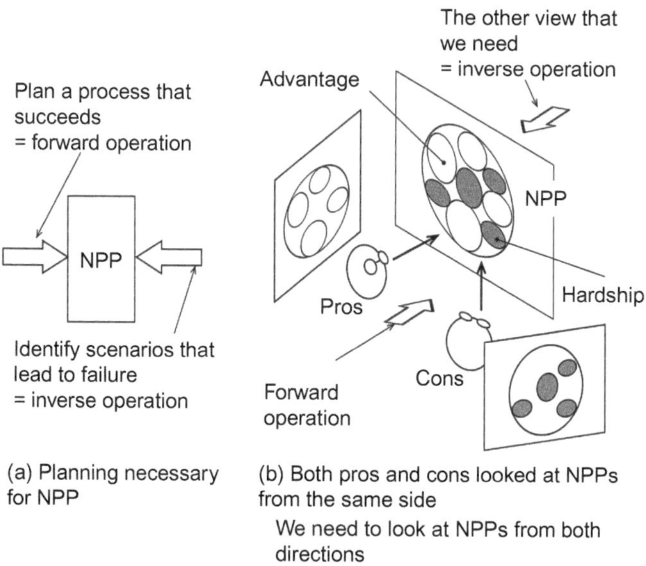

Figure 6.12 Views we need for NPPs.

in taking action against the accident. There were some mistakes and delayed responses; however, in response to the rapidly changing situation they saw with their own eyes, they made judgments about what they could prepare and perform and acted accordingly.

We have to develop as individuals who think and move actively and at the same time educate people to develop similarly.

The great work by field workers prevented the worst situation; however on the other hand, maybe this accident raised concerns about nuclear engineers in Japan. Isn't it true that the engineers had kept too silent up to this time?

Let us introduce an experience of one of the authors, Hatamura. When he visited Sandia National Laboratories in the United States, one of the presentations was made by a U.S. government investigator who came to Japan to look into the 1999 JCO criticality accident in Japan. During his investigation, he interviewed each engineer, but not a single person voiced his own thoughts or opinion. The presenter said if an engineer cannot express his own thoughts toward others, the country is not qualified to handle nuclear power.

This accident has proven that the phrase "Nuclear power is safe" was a mistake. There must have been some nuclear engineers who knew that despite the number of safety measures, they were insufficient. Perhaps, those engineers should have spoken out more loudly. Unless each engineer makes subjective judgments and makes active movements, we cannot expect to have real safety with nuclear power.

This statement is true not only for workers and engineers at nuclear plants. Each one of us has to think, judge, and act accordingly. We have to educate our people to this and nourish a culture shared with such people.

6.6 Conclusions

This chapter listed the lessons we should learn from the Fukushima accident. The first set is specific to this accident; however, the latter set of lessons applies to all industries.

It is my hope that not only engineers, but also managers, salespeople, administrators, and all people striving to bring benefit to the world while they are earning their living will learn from these lessons so we do not have to repeat mistakes of the same nature in the future.

Reference

[1] Investigation committee on the accident at the Fukushima Nuclear Power Station of Tokyo Electric Power Company, Final report. *http://www.cas.go.jp/jp/seisaku/icanps/eng/final-report.html*, 2011 (accessed 08.08.14).

The future of nuclear power generation after the Fukushima accident

7

Y. Hatamura
Hatamura Institute for the Advancement of Technology, Tokyo, Japan

7.1 Introduction

Over 3 years have passed since the 2011 Fukushima accident, and the Japanese government is now headed in the direction of restarting some of the resting nuclear power plants. Is it safe to restart them now? What are the conditions required for the restart to take place?

This chapter displays the author's thoughts on what we need to see happen before we allow the nuclear industry to light up the reactors again.

7.2 Restarting the Fukushima plant

7.2.1 The acceptance of nuclear power generation will change

Currently, more people are inclined to abandon nuclear power generation altogether. This, however, is probably due to people's minds being in a transitional state immediately after an accident.

The accident led to stopping all domestic nuclear power plants, and none of them have restarted except for the Ooi power plant. The entire country suffered electric power shortages, and our electricity bills went up. If this situation persists too long, industries, which had already been weakening before the accident, will only increase their problems to face a weaker Yen and could even reach a point where we cannot import sufficient energy and food. Can we bear such hard lives? Under the circumstances, there is a large possibility that society will change its ideas with time elapsed from the accident. When we plan the future of nuclear power generation, we also have to account for the possible change in people's opinion.

After World War II, as the Japanese economy recovered its strength, the demand for electricity soared, and its shortage was restricting economic growth and improvements in lifestyle. Kansai Electric Power Company (KEPCO) at the time poured twice its capital then to build the Kurobe River Hydraulic Power Plant Unit 4. The construction work claimed the lives of as many as 170 workers. Kurobe Unit 4 has a capacity of 335 MW,

The 2011 Fukushima Nuclear Power Plant Accident. http://dx.doi.org/10.1016/B978-0-08-100118-9.00007-3

about a quarter of that of the newest single nuclear power reactor. Japan, in the midst of power shortage in the 1960s, decided to introduce nuclear power generation to overcome the hardships. Looking back at the history of how it started makes us wonder if we are simply stepping back in time to repeat the electricity shortage again.

After 30 or so years, the generation that lived through the time of this accident will reduce to about two-thirds of the whole population, and the generation without this experience will be leading society. When members of a society change, the way people think may also change as well.

We already discussed how U.S. society shifted toward accepting nuclear power generation. We have to remember how society changes its ideas against nuclear power generation with other aspects of society at the time.

7.2.2 The cost of nuclear energy compared with other sources of energy

So far, we have heard that cost of power generation is less with nuclear compared to other sources of energy. This accident, however, revealed that it is not so. The authors estimate the losses with this accident will reach at least 50 trillion Yen including what has not been announced to the public. Estimating the total amount of electricity generated up to this point (in about 50 years) at 7.5 trillion kWh, we reach an added cost of about 7 Yen per 1 kWh due to the accident. The cost of generating electricity with nuclear power has been said to be about 5–6 Yen/kWh before the accident; however, we now have a new estimate of about 12–13 Yen/kWh. The cost of generating electricity with nuclear power is not low at all.

When we want to decide whether we are pro or con nuclear power, we must acknowledge that what was said to be safe and at low cost before is actually really dangerous and at high cost. With that, each and every citizen has to think whether we shall keep using nuclear power generation or not.

7.2.3 Alternative sources of power

Is it feasible to meet the ever-increasing demand for electrical power with fossil fuel and renewable energy? At the current state of technical resources, we cannot meet all the electricity demand Japan has with nuclear power.

Then our only option is to increase electricity production with fossil fuel as an alternative source of energy. Japan's fossil fuel supply relies almost entirely on imports and if the cost rises in the future, the cost of power generation will push electricity bills up, and the nation's weakness of having to rely on its energy sources from foreign countries will reveal more disadvantages.

Also, no matter how we improve the efficiency of producing electricity, increasing fossil fuel power means adding to carbon dioxide release in the atmosphere. We now hardly hear any concerns about the problem of carbon dioxide emission. Whatever happened to the problem of global warming?

Some may think that we have managed to survive without nuclear power after the accident; however, the fact is that we are pushing the limits of old outdated fossil fuel power facilities as emergency measures.

Others suggest following the German move of replacing nuclear power with natural energy sources, but when we listen to people in Germany, the reality is quite different from what we see in Japan.

First of all, Germany was where the iron curtain stood between the east and west during the Cold War. The people not only had to fear nuclear wars but also have sour memories from the Chernobyl accident. Chernobyl caused hot spots not only in the former Soviet Union but also in the surrounding countries. Dairy products, mushrooms, and berries still suffer from fears of radioactivity. With all these historic events behind them, the German government announced it will abandon all 17 nuclear reactors they have by the year 2020 and will rely on alternative energy sources of wind and solar. There is, however, a mountain of problems to solve for that to happen.

For example, wind power generation is primarily adopted in the northern part of Germany, and without a good way of offloading excessive power in the winter, Germany is facing conflicts with neighboring countries who have to handle the forced power rise in their electrical grids. Southern Germany, on the other hand, has adopted solar power generation but its high altitude hampers high efficiency, and many are against the higher electricity cost, thus, larger electricity bills. Even with such issues, the country will probably keep its move toward renewable energy.

We shall not simply admire the German policy of abandoning nuclear power, but understand its hardships described here and then discuss its feasibility in Japan.

7.2.4 Coping with nuclear power generation

The Fukushima accident is one of the worst failures in the history of nuclear power generation. Nuclear power itself is not evil nor is it a savior. We should be aware of its dangers and handle it with proper fear. When we have to plan whether to restart plants or to abandon them, we should not go with the mood of the time and should have thorough discussions with a complete understanding of the preceding concerns.

Whether we keep the plants closed or restart them, we need to keep the knowledge about nuclear power generation up to date to keep the nuclear power technology alive.

One of the reasons is because we have to perform the postprocessing of the Fukushima nuclear accident. Another is, even if we decide to abandon nuclear power generation, we have to handle spent fuel from nuclear power plants around the country. Furthermore, emerging and developing countries are actively starting to introduce nuclear power generation, and even if Japan decides to abandon it, when neighboring countries continue to make use of the technology, it will be questionable if Japan can carry on without the technology of nuclear power generation. Whether Japan will export its nuclear power technology or not is a separate question. The possible need for nuclear power generation after a few decades even after we abandon it now is the topic we discussed earlier.

7.2.5 Considerations in restarting nuclear power plants

The current plan is to discuss when to restart each nuclear power plant after evaluating its safety with stress tests. We are, however, concerned as to whether these tests are adequate for deciding whether to restart a plant or not. We always have oversights, and basing decisions on the same method with tighter tolerances or stricter criteria does not open our eyes to them.

Up to the point before the Fukushima accident, we aimed at building safer nuclear power plants for effective and safe electricity supply by setting standards and rules for accident prevention. The accident, however, broke out with the tsunami attack that we had failed to acknowledge as a factor.

The accident taught us the limit of forward thinking (i.e., to plan thoroughly ahead of time and prepare accordingly for effective and safe nuclear power generation).

All of us, including the utility companies, regulators, local and central government agencies and the people, have to bear in mind that we should assume that accidents happen, think what events will take place in such cases, and plan what to do to minimize the damage.

For this purpose, "disaster prevention" that merely works to prevent accidents is insufficient, and we have to plan "damage reduction" (i.e., "measures for preventing the spread of damage") with the assumption that accidents do happen.

7.2.6 Damage reduction measures required for restarting nuclear power plants

Whether we restart them or not, as long as the nuclear power plants are there, the possibility of large-scale accidents with external radiation release persists. Such accidents will affect not only the local residents but also their families, regions, and communities. The utility companies, administration, local governments, residents, and the entire nation, without recognizing this fact, allowed the building and operating of nuclear power plants, and in addition lived in their neighborhood. If we had had better understanding and preparation, we would have made totally different responses and could have avoided the large-scale catastrophe.

It was not just a problem with the local residents, but the whole nation did not have proper understanding of nuclear power generation and radioactivity, and we overreacted with fear against radioactivity and caused uncalled-for rumors that worsened the damage. We need to provide education about nuclear power generation and radioactivity from the early stages of education during the first to the ninth grades.

To restart the nuclear plants, disaster reduction, in other words, measures for preventing the spread of damage are needed, and one of the most important measures is to carry out evacuation exercises to the extent similar to the one that had to be taken with this accident.

The evacuation during this accident did not proceed smoothly. As we discussed in Chapter 3, Futaba hospital in the town of Futaba moved its hard-to-move patients without thorough plans, and a number of them died during the evacuation or immediately after settling in at the sanctuaries. Evacuation exercises with all regional residents' participation in a form close to reality are most important in planning evacuation in addition to building schedules and preparing locations. These exercises will not only teach the residents living near nuclear plants what actions to take but they will also reveal the oversights of how traffic jams occur and which roads are not really ready for evacuation.

We also have to note that the affected region is now faced with the problem of where to store the contaminated soil. As we discussed earlier, radioactive material cannot be eliminated, and we have to wait for the radioactivity to decay. Plans to move the contaminated soil from one place to another will cause objection from residents in the target area. We have to build plans ahead of time about how to handle contamination and inform the residents in the area to gain their prior understanding. The 160,000 evacuees are voicing their anxiety about their future. We have to plan not just about the evacuation immediately after the accident but also about the long-term measures of how the residents will recover their former lifestyles.

7.3 Conclusions: the future of nuclear power in Japan

In evaluating whether to restart the nuclear power plants or to abandon them, we need a thorough discussion with consideration to the future of Japan as an entire nation. We also need to nourish a culture that accepts dangerous matters as they are, and this need is not just limited to nuclear power generation. Through such discussions we must find a new balance between convenience and burden.

What is important, then, is not to rely on opinions of people around us. Each of us has to picture the overall state of Japan in the future and discuss the issue. One option is to accept the danger with it but to rely on nuclear power to secure enough power supply. Another is to abandon nuclear power generation with the risk of Japan losing its economic strength and facing energy and food shortages.

Whatever the case, each of us, each Japanese national, must make a decision for the future of the nation.

Postscript

On July 23, 2012, our investigation committee published the “Final Report on the Accident at the Fukushima Nuclear Power Stations of the Tokyo Electric Power Company”. The government’s investigation committee had completed the report at the beginning of July, however, so the committee was active for 13 months, from early July 2011 until late June 2012. The committee knew that nuclear power generation doesn’t simply raise safety concerns. Nuclear power generation affects the entire society because of its importance in providing the nation’s energy supply. Thus, the accident investigation had to meet two distinct requirements: it had to produce a report as quickly as possible, and the report had to reflect high standards of proof and reasoning. The so-called Kemeny committee, assigned by President Carter, turned in its report 6 months after the team was formed, and our committee started with a target completion time of about 1 year.

Our interim report contained about 451,800 Japanese characters, and our final report contained 182,700, summing to over 630,000 Japanese characters (roughly equivalent to 250,000 English words). The report identified the basic facts about the Fukushima nuclear power plant accident in order to achieve an acceptable grade.

Unavoidably, producing such a huge number of pages within such a short period of time led to a document that was limited in terms of content and format. Without a thorough review of the language, awkward expressions were left in many places, and the report did not even have an index. Thus, readers interested in the committee’s findings had to tackle a report that was clumsy, bulky, and difficult to read.

For this very reason, as members of the investigation team, we decided to write this book. In other words, we wanted to rewrite the essence of the two reports in plain words so more people could read about the accident.

This book is based on the committee’s interim and final reports, but we have added our personal views in the text. Also, the report of the investigation committee was an official government document, and as a result, it avoided definite conclusions and lowered tone in making evaluations. This book, in contrast, presents evaluations based on the personal opinions of the authors. Three of the authors served as members or technical consultants to the investigation committee. However, this book is not a complete recreation of the investigation report, and the four authors take responsibility for any errors that the book might contain.

As noted, the investigation report was limited in terms of the data it could consider. For example, the high radioactive doses present during the accident and its aftermath blocked entry into the reactor building, and, thus, the damage to the primary facilities, the core damage, and the leakage route of hydrogen and radioactive material remained unknown at the time the report was being prepared. Also, the committee could only use information available as of May 2012 when evaluating the accident's health effects; the potential contamination of agricultural, dairy, and fishery products; chemical and radiation levels in air, soil, and water; and even the emergency responses by local governments. In addition, the report did not discuss the history of nuclear power technology and the changes in Japanese policy regarding nuclear power. Finally, when the committee was formed, we considered investigating whether academic societies, universities, or research institutes acted in ways that contributed to the accident, and we also hoped to clarify the role of nuclear plant manufacturers in the failure of the power stations, but we had no time to look into these subjects.

In other words, the report by the investigation committee was a mere list of basic facts that formed the foundation for full analyses to be made in the future. These facts include: (1) Events and emergency responses by the Tokyo Electric Power Company (TEPCO) at the Fukushima-1 Nuclear Power Plant made during the one-month period starting on March 11; (2) How the damage expanded and the resulting resident evacuation outside the site; (3) Emergency responses by the government and problems with them; and (4) Problems with the government's nuclear safety regulations in the 1990s. Using this information, the government, TEPCO, and related academic societies and organizations should continue comprehensive and thorough investigations about the remaining problems with the Fukushima Nuclear Power Plant accident. This effort can also involve international organizations.

In the course of working on the investigation report, the committee organized a "Meeting with Invited International Experts" in Tokyo on February 24 and 25, 2012, in order to conduct an international peer review of the committee's findings. Among the comments from the foreign experts, two left strong impressions in my mind. First, experts noted the problem of workers operating on site without dosimeters. Second, André-Claude Lacoste, then-President of the French Nuclear Safety Authority (ASN), said, "I had the fear that the next large-scale nuclear accident would take place in Japan." I asked Mr Lacoste on what grounds he came to that conclusion, and he replied that he was concerned that Japanese safety regulations had fallen far behind the regulations in the rest of the world. We have to take his words quite seriously. The most important task of the newly established Nuclear Regulation Authority is to quickly improve Japanese safety regulations so that they meet international standards.

In writing this book, the authors had support from many people, including radiation protection expert Shizuko Kakinuma and the other members of the investigation committee; the Mayor of Kawamata-cho in Fukushima prefecture, Michio Furukawa a non-fiction writer, Kunio Yanagida; Shinji Ogawa of the Investigation Committee Administration office and his nearly 30 staff members; Professor Seiichi Koshizuka and other experts in the office. The staff members from the Hatamura Advanced Institute of Technology also offered great help in editing the diagrams in the book. We would like to express our gratitude to them all.

September, 2014
Seiji Abe, Kansai University, Japan

Index

Note: Page numbers followed by f indicate figures and t indicate tables.

FSC
www.fsc.org
MIX
Papier | Fördert gute Waldnutzung
FSC® C083411

Kolibri 360 GmbH
Ferdinand-Jühlke-Straße 7
99095 Erfurt, Deutschland
produktsicherheit@kolibri360.de